The Great Western Steam Retreat

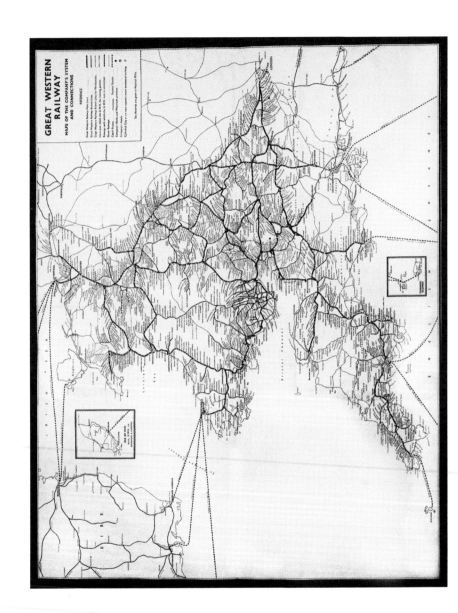

GREAT WESTERN RAILWAY (GWR) SYSTEM MAP

The Great Western Steam Retreat

Chasing the Final Steam Trains
in BR's Western Region, Wales
and the Welsh Marches

Keith Widdowson

Front cover image: 7029 *Clun Castle* at Stourbridge Junction.
Back cover image: 6916 *Misterton Hall* accelerating through Oxford.

Every effort has been made to source and contact all copyright holders of illustrated material. In case of any omission, please contact the author care of the publishers.

First published 2022

The History Press
97 St George's Place, Cheltenham,
Gloucestershire, GL50 3QB
www.thehistorypress.co.uk

British Library Cataloguing in Publication Data.
A catalogue record for this book is available from the British Library.

ISBN 978 0 7509 9807 9

Typesetting and origination by The History Press
Printed in Turkey by IMAK

Contents

About the Author

Keith Widdowson was born to a pharmacist father and secretarial mother during the calamitous winter of 1947 at St Mary Cray, Kent, and attended the nearby schools of Poverest and Charterhouse. He joined British Railways (BR) in June 1962 as an enquiry clerk at the Waterloo telephone bureau 'because his mother had noted his obsession with collecting timetables'.

Thus began a forty-five-year career within various train planning departments throughout BR, the bulk of which was at Waterloo but also included locations at Cannon Street, Wimbledon, Crewe, Euston, Blackfriars, Paddington and finally Croydon – specialising in dealing with train crew arrangements.

After spending several years during the 1970s and '80s in Cheshire, London and Sittingbourne, he returned to his roots in 1985. There he finally met the steadying influence in his life, Joan, with whom he had a daughter, Victoria. In addition to membership of the local residents' association (St Paul's Cray), the Sittingbourne & Kemsley Light Railway and the U3A organisation, he keeps busy writing articles for railway magazines and gardening.

Acknowledgements

This book is dedicated to the many people in my life who have made it one that I have been glad to have participated in – my ever-understanding wife Joan being at the top of the list. Then there is Joe Jolliffee, a lifelong fellow basher, who proofread this tome, and Chris Magner, a member of the Unofficial Volunteer Birkenhead Steam Cleaning gang, who did the same. Also John Bird (ANISTR.com), whose miracles on fifty-year-old-plus negatives have made them worthy of inclusion here. Many thanks also go to The History Press team for putting it all together, and finally there is *Steam Days* magazine editor Rex Kennedy, who in 2004 published my first article, thus kick-starting a late-life career as an author. You can view all of my photographs on www.mistermixedtraction.smugmug.com, where you can click on any of the galleries, set it up for a slide show, sit back and enjoy.

Introduction

Born at home in the winter of 1947, my childhood during the 1950s was unexceptional. Come school age, with Dad commuting to London, Mum returned to work in a nearby factory and we, my younger brother and I, became what are nowadays known as latchkey children, i.e. we looked after ourselves until a parent returned home. Fortunately, having woods opposite our house (adjacent to the Up Chatham line at St Mary Cray), many an hour was spent exploring or walking our dog (who enjoyed attempting to race London-bound trains just starting out of the station) there. Alternatively, when it was dark or there was inclement weather, there were always comics such as the *Dandy*, *Beano* or *Beezer* to read. On other occasions, we raced each other on our tricycles around the block with the instructions 'don't cross any roads' echoing in our ears – mine had solid tyres while his had the more comfortable pneumatics *and* a basket on the back to boot. The first born always loses out! Another difference was that, one Christmas, I was given a clockwork Hornby O-gauge model railway set, while he had a more modern Scalextric racing track.

Having failed the eleven-plus, I was dispatched to Charterhouse (Orpington) School, which involved a daily bus journey. With the London Transport services not being as frequent as they are nowadays, it became essential to read a timetable – with hindsight the catalyst that led me into a railway career. More of which later. Other non-school activities experienced were cubs/scouts, paper rounds (both mornings and the Saturday evening *Pink Classified*), the Youth Hostel Association and days out on the buses with our Green or Red Rovers. At home, using the money from the paper rounds, records by Cliff Richard, Adam Faith, Duane Eddy and The Shadows were purchased and played at full volume on my Dansette record player – much to my parents' annoyance. Listening to Radio Luxembourg on my Pye radio with my earphones under the bedclothes only heightened the 'illegality' of it – not forgetting the annoyingly persistent Horace Batchelor (of Keynsham, don't you know) adverts about his sure-fire method of winning a fortune on the football pools. If the scheme was that good, then why wasn't he a millionaire?

Educationally I didn't to do very well and, when I was 15, my father said to me, 'The RSA certificate, which you *might* achieve, won't get you anywhere. You may as well leave and get a job.' Mum, having noted my obsessive interest in collecting and reading timetables, wrote to the Southern Region (SR) HQ at Waterloo and obtained an interview for me in the telephone enquiry bureau, whose role was to answer enquires from the public by reading timetables. Having successfully passed an entrance exam together with a medical, I commenced what was to turn out to be a forty-five-year career within the railway industry in June 1962.

Doctor Beeching had, in March 1963, published his *Reshaping of British Rail* proposals – the crux of which foresaw widespread railway line closures and the elimination of the steam locomotive. Although only joining BR during the previous year, I don't recall having employment concerns that resulted from the report's recommendations. Perhaps the exuberance and naivety of youth shielded me from any worries regarding possible unemployment. The Southern Region, perhaps the least affected, was essentially a commuter railway and there always seemed to be a shortfall of clerks to man the phones and plenty of overtime on offer – management often called upon retired staff to fill the vacancies. I had originally joined BR 'just as a job'. However, being an impressionable teenager and urged on by fellow enthusiast clerks, namely the late Bill Sumner, to visit the destinations for which I was forever answering train enquiries, I began to tentatively venture out into the world away from my daily commute. Few records or notes were made during that period, and all I took along with me was a Brownie 127 camera and a fold-out system map (extracted from the rear of an SR timetable) of railway lines, which I coloured in as I travelled over each.

In August 1963, I holidayed with my parents and brother at Woolacombe, but perhaps showing my growing independence (at the age of 16), I travelled to and from there by train. While waiting for my returning London train at Mortehoe, 31-year-old Churchward-designed Mogul 7337, withdrawn thirteen months later at Swindon, arrived there with the 0620 Taunton to Ilfracombe. It was assisted up Mortehoe bank, parts of which were 1 in 40, by N 31842. The poor quality of this photograph can be attributed to the fact that it was taken with my Brownie 127.

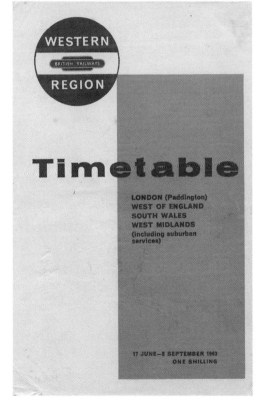

The summer 1963 Western Region (WR) timetable.

Our family holidays had nearly always centred on resorts in the west of England. I well remember the traffic queues along the A30 and the notorious Exeter bypass, with *I Spy* books, usually involving car registrations from across the country, sometimes alleviating the boredom. Perhaps indicative of my growing independence, I journeyed by train in 1963 while my parents drove as usual – my travel perks from my BR employment being an obvious bonus. I have vague recollections of being in a Bulleid compartment side-corridor coach but, as to the motive power on the 223-mile journey from Waterloo to Mortehoe, I didn't note it. Completing the few miles to Woolacombe by bus, I re-joined my family before returning a week later by the same method.

So that brings us up to 1964 – and I began to document every non-commuting journey undertaken. Rough notes were made in small notebooks, the first few of which were lost in the mists of time – not, fortuitously, before the contents were neatly transferred into A4 desk diaries. All those notebooks from 1965 have survived, the first dozen or so pages being of tidy writing, but both the condition of the book and my writing deteriorated during its usual six-month tenure because of their incessant usage during my travels.

On 1 January 1963, there were 7,074 steam locomotives in BR stock, a decrease of 1,011 since 1962. A year later, this number would be reduced to 4,990. Little did I realise this, and had I done so, I would have travelled far more extensively (funds permitting) to chase after them before it was too late. It would be easy to blame the meagre wages (£2 12s 6d per week, or £2 62½p) for the few journeys made, but in reality, it was my own naivety as to the impending massacre of Britain's Iron Horses.

The May 1964 Part 1 BR Steam Locomotives ABC.

Introduction

The lack of funds meant outings were initially confined to the SR – my home region. Gradually, as I progressed up through the clerical ranks, my increased income, coupled with qualification for an escalating number of free passes each year, meant my horizons began to expand. Promotion in 1966 allowed me to afford rail tours. The Saturday visits became overnight Friday – then Friday to Sunday – then Thursday to Sunday – eventually culminating in a five-night bash on the London Midland Region (LMR) in the autumn of 1966 – the year of my greatest ever yearly steam mileage of 35,528.

It would have been so much easier to 'cop' locomotives at stations, yards, depots, etc. It was far more difficult, and by default more rewarding, to travel behind them. The satisfaction of 'clearing' a class or shed allocation, or getting hauled by one from a predominantly freight depot or a 'namer' on an unexpected turn-up, gave teenagers such as myself a thrill and a certain kudos among like-minded contemporaries. Compared with almost completed Bulleid pages and respectable Brit and Black 5 entries, the Great Western Railway (GWR) pages were desperately bare.

Much as I enjoyed the camaraderie of steam-chasing days with fellow enthusiasts, I cherished my solo ventures too. It was just as well because on the majority of expeditions chasing after WR steam, I was on my own! The camaraderie came along upon steam's contraction to north-west England, the overnight Calder Valley mail train scenario and, of course, the Waterloo to Bournemouth main line. Even so, there were usually identifiable railway followers at the platform ends of most large stations. If information was required of them as to what had been through recently or of anything unusual happening, my questions were usually met with a positive response. Generally, the dress code gave them away – a person with a notebook in hand and with no intention of boarding any train. We came from all walks of life and there was no snobbery; all that mattered was your knowledge of the railway.

I was a child of the interregnum caught up in the uneasy period between the glorious years of the steam age and the cleaned-up Inter-City-branded future. The death of the steam locomotive meant little to us – we were young and we were going to live forever. We were oblivious, ignorant even, to the political skulduggery that was to do away with both the steam locomotive and coal fires. Being a baby boomer, I consider myself privileged to have witnessed, sad though they were, the last few years of steam on the national network. Thereafter, the future looked bleak with steam being banned throughout mainland UK and fast disappearing throughout Europe. The reign of the steam locomotive was thankfully lengthened by the decision in the 1950s to undertake a build of Standard locomotives, culminating with the construction of BR 9F 92220 *Evening Star* in March 1960 at Swindon.

To me a 'Western' was a television show or series that took place in the American old west that involved cowboys, cattle ranchers, miners, farmers, Native Americans, guns and horses. It was the most popular genre of TV show in the 1950s and '60s, and I grew up on a diet of programmes such as *The Adventures of Champion* (the wonder horse), *The Adventures of Rin Tin Tin, Rawhide, The Lone Ranger, Laramie* and *The Roy Rogers Show.* However, having become a BR employee, 'Western' took on a very different meaning. Reading the office copy of magazines such as *Railway World*, it slowly dawned on me that, with widespread closures and the hastening approach of the steam locomotives' demise, the Western Region, with no electrification plans, was selected to be the first to achieve full dieselisation.

It must have been mid-1964 when I first got wind that the WR was hell-bent on eliminating steam within the following eighteen months or so. In January 1965, WR General Manager Gerry Fiennes told *Modern Railways* he was intent on turning a £30 million loss into a break-even figure that year. He spoke of the savings possible with diesels, with operating costs falling from 4s 2d (21p)/mile to 2s 9d (14p)/mile. As a teenager living in SR territory but close enough to the final workings in the Oxford–Banbury area, I resolved (finances permitting) to record the decline and travel with some GWR examples before it was too late.

The Western Region (née Great Western Railway) was always a thorn in the side of British Railways' attempts to standardise the regions with the same lacklustre brand. The Western's history was one of past glories and lengthy family allegiances. Britain's railways had been nationalised, yet remained a loose affiliation of small republics, each one with their own identity and history. This 'loyalty' filtered down to us enthusiasts: each one of us had our favourites and we would argue their merits during the increasingly lengthy waits between steam trains.

The British Railways Board (BRB) was determined to squash this. No one had a job for life any more, and while grumbling was always part of a railwayman's resolve, at least he knew that it would blow over and, providing he kept his nose clean, there would be a pension at the end of it. Of all the pre-grouping companies, GWR remained the most geographically intact, and indeed stayed so up to and after the 1948 nationalisation. Their locomotive fleet had also been standardised since the middle of the twentieth century under the guidance of Churchward, and this was continued by his successors, Collett and Hawksworth.

There's something about the smell of coal, together with any great exhalation of steam from a locomotive, which is instantly relaxing – it is a living and breathing machine. The magical rhythmic beat of an approaching loco's exhaust sound always gave a steam follower a tingle down the back of their neck. Sheds were atmospheric cathedrals of steam and smoke – the air heavy with smoke, dust and the smell of sulphur. The ground was littered with oil-splashed puddles of black water and potential danger for those who weren't looking where they were going. We weren't bothered – it was all part and parcel of our chosen hobby. On the trains themselves, the steam heating saturated our bodies, whereas electric heating dried the back of our throats. However, now the steam locomotive, which had opened up the country during the nineteenth century, had become the victim of the relentless but necessary march of progress. To quote the prolific railway author Mike Hedderly: 'No generation of railway enthusiasts had to cram in so much in such a short time. It was a race against time.'

Self-confidence grows with age and nowadays I have no problem with declaring myself to be a railway enthusiast. Sure, I am still labelled, by some, as an anorak – so what: a hobby is a hobby and if pleasure and interest can be gained from it, so be it! These days the trainspotter has become a more acceptable persona – less marginalised because of the resurgence of interest in the steam locomotive courtesy of the *Flying Scotsman* and *Tornado*. Celebrity train enthusiasts Michael Portillo, Dan Snow and Chris Tarrant have each played their part. Spending vast amounts on photographic equipment, together with the range of mobile phones, cameras and GPS paraphernalia available to today's enthusiasts, would be beyond the understanding of those following the hobby in the 1960s.

Not discovering steam until 1964 (having had the opportunity from June 1962), I am seriously envious of those older and wiser who went before me. Even though I caught the last four years, it was the mid-1950s, the heyday of British steam, that I would have really liked to experience. If the Second World War hadn't happened (my father having been corralled in a POW camp for five years), my parents might well have processed me up to eight years earlier, so I can genuinely blame Hitler for my late arrival on the scene.

The Great Western Steam Retreat

It was like someone who arrives late at a party and is told that they should have seen what happened earlier – surveying the party food table, they are then left feeding off the leftovers.

Even these days, when the relatively rare incidence of a new haulage occurs, the hairs on the back on my neck stand up and my eyes glisten with tears at still having the ability, courtesy of hundreds of dedicated volunteers, to continue to enjoy what has been a lifetime addiction – that of journeying behind a steam locomotive I have never travelled with before, whether on a preserved line or on the national network. The thrill of underlining (or scoring through) a locomotive that I have been hauled by has never left me. When travelling the country these days and sometimes chancing on locations that once were stations or railway lines but are long since deceased, I remark that 'once upon a time' I had travelled over that line or visited the station. My wife, bless her, feigns polite interest but with disbelief that anyone can eulogise so fervently over travels made in the past.

As readers will come to appreciate in the following pages, my greatest financial outlay was on travel costs and photographs were (regrettably) of secondary importance. Thus, the majority of photos included here were taken within station limits. In this book the reader will find a series of adventures undertaken in the search for steam over routes, some of which have long gone, throughout both the former Western Region and Great Western Railway. Sit back and join me on journeys that can never be recreated.

1

Reading Reminiscences

In 1964, the world was there for a 17-year-old full of wanderlust and youthful vigour to explore. Released from the parental imposition of short back and sides, and obviously influenced by the pop groups of the day, collar-length hair and large sideboards were grown. That February, having dominated the top spot for six weeks, the Searchers' 'Needles & Pins' was finally dethroned by the Bachelors' 'Diane' – only for them to be ousted a week later by Cilla Black (on whom I had a serious crush!) singing 'Anyone Who Had a Heart'.

In my wage packet, having worked overtime during the Christmas period, I received one of the new £10 notes. With their white predecessors having been withdrawn in 1945 (to combat forgery), the new ones were not only coloured (brown) but for the first time featured a portrait of the monarch.

On Tuesday, 11 February, having taken half a day of annual leave from my workplace at Waterloo, I made my first investigative journey away from SR metals – all of 36 miles away to Reading. After crossing London on the Bakerloo line, I caught the 1230 Torbay Express departure out of Paddington, powered that day by Hydraulic D1051 *Western Ambassador*, window leaning en route in order to drink in the atmosphere of a non-electrified railway. After passing Old Oak Common and Southall depots, however, the wind resistance, together with the cold, forced me to take refuge in a window seat.

The Great Western Steam Retreat

In my notebook I penned at the time, 'Steam locomotives (predominantly Halls and Granges) were liable to work the Paddington to Worcester trains due to a series of Western diesel failures on the West of England line resulting in the transfer of Hymeks from the Worcester line.' How accurate these facts were I couldn't say; perhaps I had read them in the office copy of *Railway World*. In reality, during the time I was there the only passenger train steam substitution I witnessed was Worcester's 6856 *Stowe Grange* on the 1205 Hereford to Paddington – the same locomotive powering the 1115 out of Paddington on 8 April 1965: allegedly the final substitution.

It was, however, an interesting two-hour sojourn on the London end of Reading station and with my Brownie 127 I took some (passable?) photographs of copper-domed steam locomotives from classes never seen before. Another unique feature, witnessed for the first time, was the semaphore signals dropping (rather than raising) to indicate a clear road. This further demonstrated to a young fledgling enthusiast WR's (i.e. former GWR's) exclusivity in being different.

Guildford-allocated (although displaying a 71A Eastleigh shed code plate) Standard 4MT 80095 at Reading (Southern) in July 1963. As no notes of my pre-1964 travels have survived the years, I am unable to ascertain how I came to be there at all!

Reading Reminiscences

Oh what a fool I was! Instead of photographing Worcester's 6856 *Stowe Grange* departing Reading General with the 1205 Hereford to Paddington, I should have travelled with her. The Golden Valley trains were the last main-line services sporadically steam-hauled due to diesel locomotive (DL) shortages, predominantly Hymek.

Reading General station pilot that day was home-allocated Modified Hall 6991 *Acton Burnell Hall*. Subsequently transferred to Oxford, she made it to the bitter end in December 1965.

As per Otis Redding's 1967 hit, sitting on the dock of the bay that day was 4-6-0 7817 *Garsington Manor* on milk monitoring duties. This 25-year-old Reading-allocated Manor was named after a Tudor building 5 miles south-east of Oxford, Garsington Manor, being restored in the 1920s after a period of disrepair after having been used as a farmhouse. This milk dock was extended and became part of the terminal platform for SR services upon closure to passengers of Reading (Southern) in September 1965.

Bulleid-designed 0-6-0 Q1 33040 is seen passing through with inter-regional transfer freight to its home depot of Feltham. Shorn of all non-essentials, such as running plates, splashes and conventional boiler cladding, to save metal during their wartime construction, forty of these unique-looking 5F-classified machines were built during 1942, this one being withdrawn in June 1964.

Reading Reminiscences

Reading General, as it was named at the time of my visit, was opened from London in 1840. The line was extended through to Bristol the following year and routes to Hungerford and Basingstoke were added in 1847 and 1848 respectively. The station has been redeveloped on several occasions over the intermediate years and nowadays is a far cry from the one I first visited. Renamed Reading General in 1949 to distinguish it from the SR-operated Reading Southern, perhaps the most significant rebuild in the 1960s was in September 1965 when Reading Southern was closed and the SR EMU (electric multiple unit) trains from Waterloo and DEMUs (diesel–electric multiple units) from what is now marketed as the North Downs line were diverted into a new platform, 4A. Over the intervening years, however, it became a major bottleneck and, as a precursor to the planned electrification, it received an £897 million rebuild that included five new platforms and an additional flying crossover. It was perhaps a justifiable expense on a station that is the ninth busiest outside London. The Queen reopened the station in July 2014.

2

The First Overnight

As the pirate radio station Radio Caroline began regular broadcasts, playing hits such as Billy J. Kramer's 'Little Children' from a ship anchored outside UK territorial waters off Felixstowe in March 1964, I embarked on my very first overnight train travel en route to Wales. Little did I realise, at that time, the advantages of overnight travel, which enabled one to arrive in an area in early rather than late morning and spend a whole day exploring. Indeed, over the following four years, I spent more than 300 nights on trains or platforms throughout Britain.

My 'minder', Bill, had suggested a visit there to travel over the doomed Neath to Pontypool Road line. It was, retrospectively, his enthusiasm that perhaps provided the catalyst for my lifelong interest in railways, or more pertinently steam. Bill had planned it all, I just tagged along. Suitably laden with extra equipment not previously taken with me on daytime trips, namely toothbrush, flannel and large(r) supplies of food, I met Bill on Paddington's Lawn at just gone midnight on Saturday morning, the 21st. We were to catch the unusually routed (via Gloucester) 0100 departure for Swansea. Noise associated with long station stops while BR and General Post Office (GPO) staff offloaded papers and mail traffic meant only intermittent sleep was obtained during the six-and-a-half-hour, 216-mile journey, and Hymek diesel-hydraulic D7026 deposited us at Swansea at just gone 0730.

The First Overnight

The only steam locomotive I witnessed during the thirty-minute dwell time I had at Carmarthen on Saturday, 21 March 1964 was home-allocated 7815 *Fritwell Manor*, having worked in on that morning's 0715 from Aberystwyth. That line was proposed for closure in February 1965 but due to flood damage the northern section ceased operation two months earlier.

The 1105 out of Neath was our train over the 41¾-mile line. To kill time prior to that we journeyed along to Carmarthen, the only steam noted there being 4-6-0 7815 *Fritwell Manor* arriving on the 0715 from Aberystwyth.

The western end of line we were to travel over that day was constructed by the Vale of Neath Railway and opened in 1851. Although the Taff Vale Extension of the Newport, Abergavenny and Hereford Railway from Pontypool Road to Quakers Yard was opened in 1858, it was a further six years before the connecting link between the two railways, at Middle Duffren Junction, was accomplished. Constructed along the head of most valleys during the line's heyday, it was exceptionally busy, with the connections with numerous other valley lines contributing heavily to its usefulness. As the coal mining industry declined during the 1950s and '60s so did the line's usage and in June 1964 it closed to passengers, various small sections lingering on for freight traffic for a few more years.

Our 1105 departure from Neath was delayed by five minutes, with preference being given to several relief trains run in connection with the rugby at

Cardiff. Collett 0-6-2T 5659 was to be our steed for the Pontypool Road train and what a wonderful experience that journey was. Initially traversing lush green valleys, the landscape was to give way to hills adorned with all the paraphernalia associated with the vast coal mining industry, some of which were scarred by vast quantities of slag. In the far distance snow-capped mountains could be seen, and a wide-eyed 'Southerner' was riveted to scenes he'd never witnessed before. The highlight, of course, was Britain's highest railway viaduct – the 200ft-high Crumlin viaduct – crossed at a speed-restricted 20mph. Its maintenance costs contributed to the line's closure that June.

Table 137— *continued*

SWANSEA, NEATH, ABERDARE, CARDIFF and PONTYPOOL ROAD

WEEK DAYS ONLY

Miles		am B		am B		am B		am		am		am		am		noon B		am E	am S	pm E B		pm S B
—	**104Swansea** (H. St.) .. dep	..	..	..	..	..	..	..	..	..	..	6 55	..	..	..	..	..	10 30	10 30	..	..	..
7¾	**Neath** (General) .... dep											7 40						11 7	11 5			
9½	Aberdylais Halt											7 45						11 12	11 10			
11¾	Clyne Halt																					
12¾	Melyncourt Halt																					
13¾	Resolven											7 53						11 20	11 19			
16¾	Glyn Neath											8 0						11 27	11 27			
18½	Pontwalby Halt											8 5										
21¼	Rhigos Halt																					
23¼	Hirwaun .... arr											8 19						11 44	11 44			
	Hirwaun dep											8 25						11 49	11 49			
25¾	Trecynon Halt											8 30						11 55	11 55			
26¾	**Aberdare** (H.L.) .. arr											**8 33**						**12 0**	**12 0**			
	Aberdare (H.L.) dep			**6 24**				**7 45**		**8 5**		**8 35**		**11 0**				**12 2**	**12 2**			
28¼	Cwmbach Halt			6 28				7 48		8 8		8 39		11 3				12 6	12 6			
30½	Mountain Ash (Cardiff Rd.)			6 32				7 55		8 14		8 44		11 9				12 11	12 11			
31½	Penrhiwceiber (H.L.)			6 36				7 59		8 17		8 48		1113				12 14	12 14			
33½	**Quaker's Yd.** (H.L.) **G** arr			**6 41**				**8 4**		**8 23**		**8 53**		**1120**				**12 20**	**12 20**			
—	**125**Merthyr ¶ .... dep	Dep Dowlais (Cae Harris) 5 20 am (Table 139)		5 55		Dep Dowlais (Cae Harris) 7 38 am (Table 139)		7 25		8 0		8 23		11 0		Dep Dowlais (Cae Harris) 11 32 am (Table 139)		12 0	12 0	Dep Dowlais (Cae Harris) 1 15 pm (Table 139)		Dep Dowlais (Cae Harris) 1 55 pm (Table 139)
—	**Quaker's Yd.** (H.L.). dep			**6 42**				**8 6**		**8 24**		**8 54**		**1121**				**12 21**	**12 21**			
34½	Treharris			6 44				8 9		8 27		8 56		1124				12 24	12 24			
34¾	Trelewis Halt			6 47				8 12		8 30		8 59		1127				12 27	12 27			
35½	**Nelson and Llancaiach** .. arr	**5 49**		**6 49**		**8 6**		**8 14**		**8 32**		**9 1**		**1130**		**12 0**		**12 29**	**12 29**	**1 48**		**2 20**
	Nelson and Llancaiach dep	**5 54**		**6 50**		**8 7**		**8 15**		**8 33**		**9 2**		**1131**		**12 2**		**12 30**	**12 30**			**2 22**
38½	Hengoed (H.L.) **D**.... arr	6 0				8 13		8 22				9 8		1137				12 37	12 37			
—	Mls — Hengoed (L.L.) .. dep	6 4				8 17						1013		1213				12 53	12 53			
—	Mls 1 Ystrad Mynach	6 7		6 55		8 20				8 39		1016		1217		12 7		12 56	12 56			2 27
—	Mls 13½ **Cardiff** (Q. St.) .. arr	**6 38**		**7‡30**		**8 51**				**9 14**		**1047**		**1248**		**12‡48**		**1 28**	**1 28**	Workmen's Train		**3‡ 6**
—	Hengoed (H.L.) ...... dep	Workmen's Train						8 30				9 9		1140				12 38	12 38			
40½	Pontllanfraith (L.L.)							8 34				9 13		1145				12 42	12 42			
41¾	Pentwynmawr Platform																					
43¼	Crumlin (H.L.)							8 42				9 21		1152				12 50	12 50			
44	Hafodyrynys Platform							8 48										12 56	12 56			
48½	Pontypool **A**							9 0				9 39		1211				1 10	1 12			
49½	**Pontypool Road**.... arr							**9 6**				**9 44**		**1215**				**1 14**	**1 19**			
59½	**164**Newport .. .. arr		..	..	..	..	..	9 42	..	..	..	11e8	..	1258	..	..	..	2 2	1 58	..	..	..

For Notes, see page 414

Final timetable extract of the Vale of Neath line.

The First Overnight

Pontypool Road's 0-6-2T 5659 pauses at Hirwaun with the 11.05 Neath General to Pontypool Road, 23 miles into her two hour, 41¾-mile journey. Between 1924 and 1928, 200 of these 5MT Collett-designed tanks were built, with this one ending her days at Croes Newydd in November 1965.

Taken by leaning out of the train window (hence the poor quality!), we are now approaching the 200ft-high Crumlin viaduct. Completed in 1857 to cross the River Ebbw, this 1,650ft-long railway viaduct was the highest in the UK. Following the line's closure in June 1964, it was dismantled three years later.

Two and a quarter hours later, at 1320, we finally arrived at Pontypool Road and headed the short distance to Newport, before retracing our outward route all the way back via Gloucester to the Golden Valley junction station of Kemble. We were here that evening because the two branch lines, to Tetbury and Cirencester, were to close the following month.

Time for some further brief history, methinks. The 1843-built Swindon to Cheltenham line opened with the provision of a station at Tetbury Road. As this was 7 miles from Tetbury itself, the town's residents felt they were losing out on the benefits rail connection would bring and proposed a connecting railway be built in 1863. It was to be an astonishing twenty-six years, amid contractors going broke and financial difficulties, before the line was opened.

In 1959, railcars replaced the steam services in a bid to save costs and two extra halts were constructed, one of which was at Trouble House Halt. This in itself was unique in that it was one of only two, the other being Berney Arms (Norfolk), where stations were specifically opened to serve public houses. Dieselisation, however, failed to stem the mounting losses and on the last day in April 1964, a coffin was loaded on the final train at the

My first sighting of a Castle was Hereford's 5042 *Winchester Castle* at Pontypool Road. Having been withdrawn and reinstated twice during her twenty-nine-year life, this then Cardiff East Dock-allocated 4-6-0 was to meet with her maker at Gloucester in June 1965.

The First Overnight

Another first was the sighting of a Franco–Crosti-boilered BR 8F 2-10-0, 92029, seen passing through Gloucester Central with an eastbound freight. This youthful 8-year-old Birkenhead-allocated machine was withdrawn upon the shed's closure in November 1967.

In fading light, 1959-built AC four-wheel railbus W79977 stands at Tetbury having arrived with the 1700 from Kemble. Three of these forty-six-seat AC Cars-built units had replaced steam on both this and the Cirencester branch; the hoped-for financial savings, however, failed to save them and both lines closed two weeks later.

aforementioned halt. It was addressed to Dr Beeching, to be forwarded from Kemble to Paddington. The train was delayed because of protesters placing burning hay bales on the line.

Dusk was now falling as we boarded the 1700 Tetbury branch departure. This was the first time I had travelled in one of the German-built railbuses and the bouncing, wheel screeching and shuddering vibrations while

Tables 106–107

Table I06   **KEMBLE and CIRENCESTER**

(Second class only)

[Timetable Table 106: Kemble and Cirencester, Week Days and Sundays, showing services between London (Pad.), Kemble, Park Leaze Halt, Chesterton Lane Halt and Cirencester (Town).]

A	On Saturdays arr 10 31 am	E	Except Saturdays	P	On Saturdays arr 10 24 pm
B	Bristol Omnibus Co. Ltd., Motor Services	F	On Saturdays arr 7 0 pm	S	Saturdays only
C	On Saturdays arr 10 50 am	G	On Saturdays arr 12 50 pm		
		H	On Saturdays arr 3 0 pm		

Table I07   **KEMBLE and TETBURY**

WEEK DAYS ONLY—(Second class only)

[Timetable Table 107: Kemble and Tetbury, Week Days only, showing services between London (Pad.), Kemble, Rodmarton Platform, Church's Hill Halt, Culkerton Halt, Trouble House Halt and Tetbury.]

E Except Saturdays   S Saturdays only

The two Kemble branch timetables.

rounding corners had to be experienced to be believed. After a lengthy thirty-four-minute turnaround, we returned to Kemble and then crossed over the station footbridge to join the Cirencester train, which was to depart from its dedicated bay. This branch was a far easier build. Proposed in 1836 and opened five years later, the terminus station at Cirencester had the word 'Town' added to it in 1924 to differentiate it from the Midland & South Western Junction Railway (M&SWJR) station of Cirencester Watermoor.

In contrast to our Tetbury stop, the turnaround at Cirencester was a mere six minutes and we were soon heading back from Kemble to London behind our third Hymek of the day, D7040, our mission completed. We arrived into Paddington at a not unreasonable time of 2100. Having experienced the advantage of overnight travel, I decided that *this* was the way forward with my newly realised hobby: to visit parts of Britain hitherto unexplored by me.

Upon the branch's closure the railbuses were dispatched to Yeovil, replacing steam on the Junction shuttle from January 1965.

3

Beeching Axe Victims

*T*he *Reshaping of British Railways* plan was published in March 1963 and the author Dr Beeching's brief (from the pro-road Minister of Transport Ernest Marples) was to put the railways on a profitable footing. This edict, coupled with the Western Region's acquisition of all the former Southern Region lines west of Salisbury under the boundary changes of 1 January 1963, was sufficient to kick-start me, albeit belatedly not until spring 1964, into action to travel over many of the inherited doomed lines.

It was the final saturday of April and, although my main thrust was to travel over two Hampshire lines that were destined to close a week later, I did, about midday, stray onto WR territory: thus it qualifies for a mention here. And so, the early hours of Saturday the 25th witnessed me boarding the 0115 West of England newspaper train out of Waterloo, Exmouth Junction-allocated BoB Light Pacific 34096 *Trevone* being in charge.

Changing at Salisbury onto a 0317 Weymouth (via Wimborne) departure, and after backtracking all the way via West Moors to Brockenhurst, it was the late morning 1140 Bournemouth West to Bristol Temple Meads train that saw me cross the regional boundary north of Blandford Forum, the subsequently preserved 4F 0-6-0 44422 easily coping with the four-coach, one-van load.

My first, of an eventual four, visits to the S&D was in April 1964 on the 1140 Bournemouth West to Bristol Temple Meads, seen here departing Evercreech Junction with subsequently preserved LMS 4F 0-6-0 44422.

The Somerset & Dorset Joint Railway (S&D) had eventually opened throughout in 1863 and, with the original main line being that which crossed the Somerset Levels to Highbridge, I changed at Evercreech Junction onto the connecting 1315 departure for the Bristol Channel port. The Poet Laureate John Betjeman once eulogised his journey along this route as 'idyllic and nostalgic', words that don't sit well with the accountants, and it was plain to see the WR had abandoned any hope of longevity for the system. The Collett-designed 0-6-0, 2218, minus her brass cabside numberplates as well as her smokebox equivalent, just dawdled along, taking just under an hour for the 22½-mile journey with her one-coach, one-van train calling at all stations whether there were any passengers or not. At the intermediate stations, those that were staffed, time was passed by the train crew and porters exchanging pleasantries about their families, resulting in a loss of eleven minutes en route. With its country stations and remote junctions, if ever there was a perfect line to remember years later, this was it – a throwback to a stress-free life long since disappeared. If only I had appreciated it back then. I wasn't to know that ahead of me lay more than forty years of London commuting with all its attendant trials and tribulations! For sure, and perhaps bucking against the trend, my travels throughout the mid-1960s meant my teenage years rather than my schooldays were the best years of my life.

It was then up to Taunton behind a Hymek DL where, after a fifty-minute sojourn, Yeovil-allocated U 31792 worked that day's 1625 departure to her home depot. This 20-mile line, opened by the Bristol & Exeter Railway in 1853, was Yeovil's first railway and was constructed across a flood plain – the Rivers Parrett and Yeo never being far away. With its stations some distance from the villages they purported to serve, it was inevitable that any prospective passengers found the frequent local bus services more convenient; closure was just weeks away. Part of the track bed has subsequently been used for the A3088, a connecting road from Yeovil to the A303, the popular dual-carriage route to the west of England.

Sleep deprivation had finally caught up with me and, having missed alighting at Yeovil's Town station for the shuttle to the main-line station of Yeovil Junction, I found myself being shaken awake at the train's terminus of Yeovil's Pen Mill station. Not to worry, and just eight minutes after arrival I was backtracked to the town station courtesy of BR 3MT 82044 on a 1745 Taunton departure. I then returned to my planned itinerary and enjoyed a run with the subsequently preserved 0-6-0PT 6412 on the 1750 'rail-motor'

We now move on to Taunton, where we see 2-6-0 Maunsell U 31792 preparing to work that day's 1625 departure via Langport and Martock to her home depot of Yeovil.

Table 80

TAUNTON, DURSTON and YEOVIL
WEEK DAYS ONLY

Miles			am		am	am	am	am	am	pm	pm	pm	pm		pm	pm	pm	pm		pm	pm	pm	
						S	E	E	S	S	E	S	E		S②		S	E				F	
	Taunton	dep	6 45	..	..	9 10	9 45	..	..	1240	1258	..	..	..	2 0	..	4 15	4 25	..	5 55	8 20	..	..
2¾	Creech St. Michael Halt		6 50	....	....	....	....	....	....	....	....	....	....	....	2 5	....	4 20	....	....	6 2	....	....	....
5½	Durston		6 57	..	..	9 20	..	..	..	..	..	..	..	..	2 11	..	4 26	4 34	..	6 9	8 29	..	..
7½	Lyng Halt		7 2	....	....	9 24	....	....	....	....	....	....	....	....	2 15	....	4 31	4 38	....	6 13	....	....	....
8	Athelney		7 5	..	..	9 27	9 57	..	..	1252	1 10	..	..	..	2 20	..	4 37	4 40	..	6 17	8 34	..	..
13	Langport West		7 18	....	....	9 37	10 7	....	....	1 2	1 20	....	....	....	2 29	....	4 56	4 51	....	6 26	8 43	....	....
15¼	Thorney & Kingsbury Halt		7 23	..	..	9 42	1012	..	..	1 7	1 25	..	..	..	2 36	..	5 1	4 57	..	6 31	8 48	..	..
18	Martock		7 29	....	....	9 48	1022	....	....	1 13	1 31	....	....	....	2 42	....	5 7	5 4	....	6 37	8 54	....	....
20½	Montacute		7 37	..	..	9 55	1029	..	..	1 19	1 37	..	..	..	2 48	..	5 14	5 13	..	6 43	9 0	..	..
24½	Hendford Halt		7 45	....	....	10 3	1037	....	....	1 27	1 45	....	....	....	2 56	....	5 22	5 21	....	6 51	9 8	....	..
25½	Yeovil Town	arr	7 48	..	..	10 6	1040	..	..	1 30	1 48	..	..	..	2 59	..	5 25	5 24	..	6 54	9 11	..	....
	Yeovil Town	dep	7 50	....	8 42	10 9	1042	1131	1148	1 33	1 51	2 24	2 39	....	3 1	4 28	5 33	5 33	....	6 56	9 13	9 50	..
26	Yeovil Pen Mill	arr	7 54	..	8 46	1013	1045	1133	1150	1 37	1 55	2 26	2 41	..	3 5	4 30	5 37	5 37	..	7 0	9 16	9 52	....

Miles			am	am	am	am	am	am		am	am	pm	pm	pm	pm	pm		pm	pm	pm			
					S	E	S	E		E	S	S	S	E	S②	E				F			
	Yeovil Pen Mill	dep	7 5	8 50	9 25	9 56	1043	1052	..	1121	1132	1250	2 10	2 25	3 50	4 0	..	5 45	7 50	9 58	..	..	..
	Yeovil Town	arr	7 7	8 52	9 27	9 58	1045	1054	....	1123	1134	1252	2 12	2 27	3 52	4 2	....	5 47	7 52	10 0	....	....	....
½	Yeovil Town	dep	7 10	..	9 29	10 0	..	..	..	1124	1135	1254	..	..	3 58	4 8	..	5 50	7 54	..	..	..	..
1½	Hendford Halt		7 15	....	9 33	10 3	....	....	....	1127	1138	1258	....	....	4 2	4 12	....	5 54	7 58	....	....	....	....
5½	Montacute		7 24	..	9 42	1012	..	..	..	1136	1147	1 7	..	..	4 11	4 21	..	6 4	8 7	..	..	..	..
8	Martock		7 32	....	9 49	1019	....	....	....	1142	1152	1 13	....	....	4 17	4 28	....	6 10	8 13	....	....	....	....
10¾	Thorney & Kingsbury Halt		7 39	..	9 56	1025	..	..	..	1148	1158	1 21	..	..	4 24	4 34	..	6 16	8 19	..	..	..	..
13	Langport West		7 45	....	10 1	1030	....	....	....	1157	12 3	1 28	....	....	4 29	4 39	....	6 21	8 25	....	....	....	....
18	Athelney		7 54	..	1010	1039	..	..	..	12 6	1212	1 37	..	..	4 38	4 48	..	6 31	8 35	..	..	..	..
18½	Lyng Halt		7 57	....	1012	....	....	....	....	12 8	1214	....	....	....	4 40	4 50	....	6 33	8 37	....	....	....	....
20½	Durston		8 2	..	1017	..	..	..	..	1213	1219	..	..	..	4 45	4 55	..	6 38	8 42	..	..	..	..
23¼	Creech St. Michael Halt		8 8	....	....	....	....	....	....	1219	1225	....	....	....	4 51	5 1	....	6 44	8 48	....	....	....	....
26	Taunton	arr	8 14	..	1033	1052	..	..	..	1225	1231	1 51	..	..	5 2	5 7	..	6 51	8 55	..	..	..	..

E Except Saturdays   F Fridays and Saturdays only   S Saturdays only   ② Second class only

The final timetable of the Taunton to Yeovil line.

departure for the Junction – a service that was withdrawn in October 1966. After catching a run with Exmouth Junction's Unmodified Pacific 34084 *253 Squadron* on a stopping service to Templecombe, I returned to London with *Trevone*, the very locomotive with which I had started this convoluted itinerary earlier that morning. An unrepeatable tour, over lines subsequently closed behind steam locomotives, all bar two of which were scrapped.

Football has always had a substantial following and so there is now a brief interlude to list that year's major achievements: the First Division champions were Liverpool and the Second Division winners were Leeds United, while West Ham beat Preston North End 3-2 in the FA Cup final.

Moving on to May, Tuesday the 26th to be precise, and with Cilla Black's 'You're My World' having been ousted by the Four Pennies' 'Juliet' off the top spot, I visited another doomed line, namely the 6¼-mile branch from West Drayton to Staines (West). Before embarking on that, however, I headed to Guildford for a ride on the WR-resourced, from Redhill, 1045 from Tonbridge. With the Saturday equivalent being a Hymek DL, it was necessary to take a valuable half-day's leave if I was to attain a run with one of the fast-disappearing ex GWR locomotives. Fortune shone upon me that day, with Reading's 7813 *Freshford Manor* performing the honours.

As previously mentioned, I had begun to colour in a railway system map showing which lines I had travelled over and the objective that day was to

As part of a convoluted plan compiled to encompass travelling over both Windsor branches and several Surrey and Middlesex suburban routes, I found myself at Staines West on Tuesday, 26 May 1964 ready to travel over the 6¼-mile branch to the WR main line at West Drayton. AEC-powered Pressed Steel single railcar W55024 formed the 1528 departure that day. The line closed in March 1965.

fill in some missing gaps in Surrey and Middlesex. I thus made my way to Staines and, having walked over from the SR Central station, I caught the 1528 departure out of the WR Staines West terminus for West Drayton.

Because the area the line ran through was sparsely populated, passenger expectations never fully materialised. Built on the River Colne flood plain and surrounded by both the Staines and Wraysbury reservoirs, it was proposed for closure under the 1963 Beeching Axe, the chop finally coming in March 1965. Ironically, although the line's original promoter's desire to connect into the London & South Western Railway (L&SWR) system at Staines was rebuffed at the time (1885), after the route was severed by the construction of the M25 in 1981, a spur was built by the SR to serve the oil terminal sited in the original Staines West goods yard. At least the northern 3 miles, to Colnbrook, is still in use as both an aggregate and oil terminal for Heathrow airport. While passenger usage was scarce at the time of the line's closure, I am certain, taking into consideration London's inexorable growth, that it could have formed part of a useful outer suburban system.

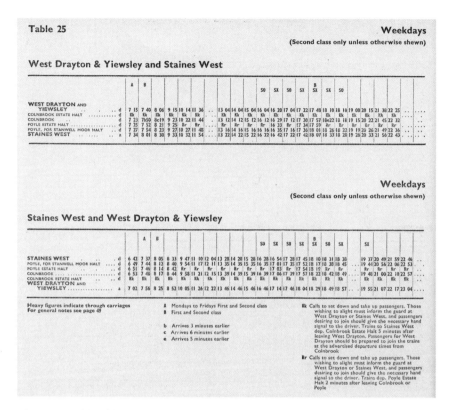

Table 25   **Weekdays**
(Second class only unless otherwise shewn)

West Drayton & Yiewsley and Staines West

Weekdays
(Second class only unless otherwise shewn)

Staines West and West Drayton & Yiewsley

Heavy figures indicate through carriages
For general notes see page 49

A   Mondays to Fridays First and Second class
B   First and Second class

b   Arrives 3 minutes earlier
c   Arrives 6 minutes earlier
e   Arrives 5 minutes earlier

Kk Calls to set down and take up passengers. Those wishing to alight must inform the guard at West Drayton or Staines West, and passengers desiring to join should give the necessary hand signal to the driver. Trains to Staines West dep. Colnbrook Estate Halt 5 minutes after leaving West Drayton. Passengers for West Drayton should be prepared to join the trains at the advertised departure times from Colnbrook

Rr Calls to set down and take up passengers. Those wishing to alight must inform the guard at West Drayton or Staines West, and passengers desiring to join should give the necessary hand signal to the driver. Trains dep. Poyle Estate Halt 2 minutes after leaving Colnbrook or Poyle

Above: Timetable of the Staines West branch.
Below: Having been ejected from Southall shed by an irate foreman, I sneaked this photograph of one of the few remaining 61xxs, once a predominate class there, in steam.

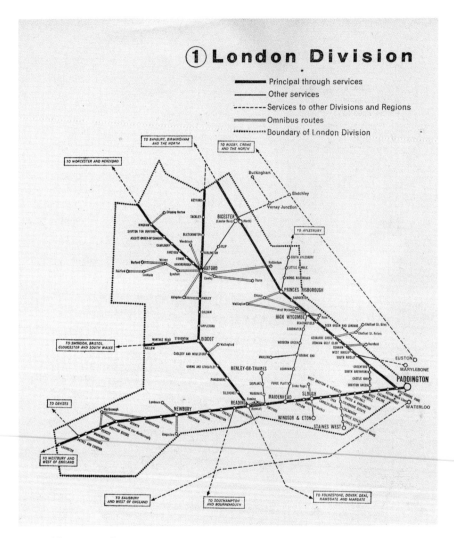

Map of the WR London Division.

Further travels that day, involving a total of sixteen different trains, enabled me to cover both Windsor branches and the Sturt Lane Chord (Brookwood to Frimley), services over the latter ceasing that September – thus resulting in a considerable amount of map colouring upon my return home.

Another London suburban visit was made in August 1965, predominantly to photograph the remaining Maunsell S15s at Feltham, after which I attempted, but failed, to get past the foreman to bunk Southall (81C) shed. En route back out of the shed, having noted the preserved 4079 *Pendennis Castle*, I did manage to snatch a shot of one of the resident 61xx locomotives. With her bunker coaled up, it was likely to have been an active example, unlike her many condemned sisters seen there. Making the most of the day, I headed to Harrow-on-the-Hill for a ride on a Great Central Railway (GCR) line steam train to Aylesbury, returning over required track via High Wycombe and noting there the subsequently preserved 2-6-2T 6106 on station pilot duties.

Three days later, the new corporate image of British Railways, or British Rail as it would become, was launched when new blue-and-grey-liveried XP64 prototype coaches were given two press runs between Marylebone and High Wycombe behind re-liveried D1733.

4

Withered Arm
Wanderings

Into July 1964 and the first number one of the year for those 'outrageous' (my parents' description!) Rolling Stones. 'It's All Over Now' was being played on my Dansette in my bedroom at full blast – enough to cause my parents to remonstrate with me to 'turn it down' from downstairs. They hated this unkempt rebellious group – perhaps as a teenager that very fact fostered a lifetime liking of their music. Incredibly, as I pen this chapter (June 2018 while in the Isle of Wight), they are *still* performing – their UK-wide 'No Filter' tour headlining at nearby Southampton.

The term 'Withered Arm' was a sobriquet railwaymen gave the former L&SWR routes west of Exeter. This referred to how the lines appeared on a map – a sparse network straddling Devon and Cornwall with a single main line splitting into a series of long, wandering branches resembling a withered limb and fingers.

As part of the 1963 regional boundary changes, the WR had acquired *all* of the ex SR lines west of Salisbury and, with their obvious preference weighted in favour of their existing main line via Taunton, the writing was on the wall for the former L&SWR route. As for the lines into Cornwall and north Devon, while the holiday resorts of Padstow, Bude and Ilfracombe attracted reasonable levels of passengers during July and August, usage during the remaining months was dire. Serving sparsely populated rural areas en route, the double whammy of increasing car ownership and an improving road network was to deal a fatal blow to their longevity. Another contributing factor was that steam, anathema to the WR HQ, still reigned supreme on all the lines involved.

With the announcement that the Waterloo to Exeter trains were to be turned over to a two-hourly, semi-fast, DL-hauled service, including the

The front cover of the WR summer 1964 timetable.

cessation of trains such as the multi-portioned Atlantic Coast Express, in September 1964, the decision was made for me: a visit to the threatened lines was a 'must do before it's too late' scenario.

Appreciating that I have extensively detailed the two journeys I made in July 1964 in my 2013 book *The Great Steam Chase*, I will just summarise the highlights here, accompanied by previously unpublished photographs, for continuity purposes as they were part of the WR.

Map of the WR Plymouth Division.

Withered Arm Wanderings

Using the twenty-four-hour clock face.

On Saturday, the 11th I travelled the 260 miles to Padstow on the 0045 out of Waterloo, the most memorable aspect being the final 50 miles through the bucolic Cornish countryside, calling at all the intermediate wayside stations aboard a two-coach portion hauled by Unmodified Bulleid 34054 *Lord Beaverbrook*. It was a typically wet summer's day and, rather than retrace my steps back to Exeter en route to my next objective of the three east Devon branches, I fortuitously made my way directly to the Great Western Main Line (GWML) station of Bodmin Road. I say fortuitously because at the time I was unaware that the 16½-mile line between Padstow and Bodmin Road was also under threat – it was just a different route on which to return back east. The 0812 from Padstow had a 4-year-old NBL *Baby Warship* Type-2 diesel-hydraulic (later designated Class 22) D6323 at its head and, after running round at the intermediate Bodmin General terminal, it deposited me at Bodmin Road at 0910. Although the line was closed to passengers in January 1967, the Wenford Bridge china clay traffic continued for a further eleven years. All is not lost, however, with the Bodmin & Wenford Railway reopening the route to Boscarne Junction (where the former SR line to Okehampton was sited) for all and sundry to enjoy.

The then Minister of Transport Ernest Marples had stated that no seaside branch lines were to close that summer – thus allowing prospective holiday-makers to plan their rail-orientated breaks without fear of luggage problems on any substitute buses. Not necessarily trusting this pro-road politician, I made my way to Exmouth by changing trains at the Exeter stations of St Davids & Central.

Travelling out of Exmouth on the one remaining loco-hauled 1334 departure with 83D's BR 3MT 82042, I then station hopped along the former L&SWR main line, calling in on both Seaton and Lyme Regis branches before heading home to the 'smoke' behind BoB 34052 *Lord Dowding*.

The Great Western Steam Retreat

It's Saturday, 11 July 1964 and, having travelled overnight from Waterloo to Padstow, I then took the 0812 Bodmin Road train, espying, at Wadebridge, a diminutive pannier tank simmering in the sidings. It was one of the six-strong Collett 1F 0-6-0PTs, 30-year-old 1369, which had arrived there in 1962 from duties on the Weymouth Quay branch, itself displacing the Beattie tanks on the weight-restricted Wenford Bridge china clay branch. Although withdrawn that autumn, she survives today on the South Devon Railway.

New to Ramsgate in 1948, Brighton-built BoB 34080 74 Squadron eases her way through Exeter St Davids with the Surbiton to Okehampton car-carrier. This then Exmouth Junction Unmodified succumbed to withdrawal in the September cull of 1964.

Withered Arm Wanderings

Swindon-built BR 3MT 2-6-2T 82042 stands at Exmouth with the 1334 (portion) for Waterloo that, at Sidmouth Junction, would combine with the 1402 from Sidmouth. Included within the 1952 production programme, it was the intention to build sixty-three of these 3MTs but Nos 82045–62 were cancelled as permanent-way and bridge improvements meant that the larger 4MTs 2-6-4Ts were able to run over routes previously earmarked for the 3MTs. This Exmouth Junction tank would be transferred to Gloucester upon 83D's closure to steam in May 1965.

Sidmouth Junction sees Merchant Navy 35019 *French Line CGT* arriving with the 1148 Plymouth to Waterloo. This then Nine Elms locomotive was withdrawn at Weymouth in September 1965.

One of the three Exmouth Junction BR 4MTs, 75022, rests at Seaton Junction. This class of eighty locomotives were all built at Swindon, with 75022 always allocated to WR sheds since being delivered to Cardiff Canton when new in December 1953. With the BR green livery being applied during her June 1962 visit to Swindon works, upon 83D's closure to steam in June 1965 she was transferred to her final shed of Worcester and was withdrawn there in the WR cull of December 1965. The station fared little better, closing, together with the Seaton branch itself, in March 1966.

Home-allocated 3MT 3210 rests adjacent to the S&D platform in the evening sun at Templecombe. Closed concurrently with the Bournemouth to Bath line in March 1966, successful petitioning by local residents saw the station reopen in 1983.

Withered Arm Wanderings

The following Saturday, with the Beatles' 'Hard Day's Night' now in situ at the top of the charts, I once again travelled out of Waterloo on the 0045 departure, this time alighting at Okehampton to travel over the Bude branch. Mission accomplished, I returned to Halwill and sat on the embankment adjacent to the single short platform from which the one morning Torrington-bound train over the North Cornwall & Devon Junction Light Railway 20-mile line was to depart. All these years later, I still recall being a teenager unfettered with any money or mortgage worries and just idling away the time that day. The sun was shining, and with just the occasional birdsong and cows mooing penetrating the lengthy silence between train movements, I considered myself fortunate in having the ability to travel the country following my hobby.

Eventually, in the distance an Unmodified Light Pacific came up from Wadebridge and came to a halt in the station; a second loco propelled the Bude 'portion' quickly and efficiently onto the rear, the entire train then getting on its way to Exeter and presumably London.

So now to the highlight of the day – the 'Mickey'-powered one-coach 1052 departure for Torrington. Opened as late as 1925, this line was constructed essentially for the china clay traffic at Meeth (which kept the northern section of the line open for freight a further fifteen years after passenger services were withdrawn) and, being classified as a light railway, was subject to a 20mph maximum speed limit, thus explaining its one-and-a-half-hour schedule. With just two passenger trains per day, sometimes becoming mixed (i.e. freight wagons attached) en route, it was hardly likely that the line was an attractive proposition to any prospective passengers. Optimistically built to 'open up the area to tourism and help both farmers and the china clay industry', most inhabitants of the largest town en route, Hatherleigh, who wanted to get to Okehampton, which was 7 miles by road and 20 by the new railway, opted for the more direct route. Conspicuously, I was the only passenger, and with the guard on several occasions having to alight to walk forward to open the crossing gates and then re-join the train after passing over the road, I felt quite exhausted just watching him. Upon arrival into Torrington the one-coach SR dual-classified carriage was exchanged for a WR two-coach set for the remaining 14 miles to north Devon's most populous town of Barnstaple.

A week later and on Saturday, 18 July 1964 N 31855 waits in the sidings for the road to the station at Okehampton to work the 0625 for Padstow. She was withdrawn in the mass cull of ex SR locomotives at Exmouth Junction, the fleet of Ns being cut from fifty-three to twelve. The survivors were all allocated at Guildford.

Halwill and an unidentified Unmodified Light Pacific arrives with the 0830 Padstow to Waterloo. Waiting to attach to the rear is the 0930 portion from Bude. Halwill was a purpose-built railway junction station and came alive every summer Saturday until September 1964. Dieselisation of all services in the form of DMUs (diesel multiple units) was effective from that date, with all lines closing in October 1966.

Perfectly illustrating how the railways were back then – Mickey 41249 rests from her exertions over the 20½-mile, one-and-a-half-hour journey at Torrington.

The next objective was the Ilfracombe branch and, although all trains to the north Devon resort appeared to have suffered from an infestation of Hymek DLs, at least my northbound train was noisily assisted up the 1 in 40 Mortehoe bank by N 31875. Returning to Barnstaple, I then boarded the 1550 Taunton departure, which was in the hands of decrepit-looking Churchward Mogul 7303. I had to ask the driver its identification as both smokebox and cabside numbers were missing; he eventually located a chalked reference *inside* the cab!

This Devon & Somerset Railway line was built in 1873, converted from broad gauge in 1881 and absorbed by GWR in 1901. Four years later, a spur was constructed at Barnstaple to allow services to access the SR line in order to run through trains to Ilfracombe from Wales and the Midlands, the GWR Victoria Road station at Barnstaple subsequently closing in 1960. This was a wonderfully scenic line crossing Exmoor and had several serious gradients, four tunnels and two viaducts to negotiate – a Victorian engineering marvel. Towards its later years, with the stations some distance from the towns and villages they purported to serve, it was only the seasonal through holiday trains that postponed the inevitable closure. My nearly two-hour, 46-mile ride that day was somewhat uncomfortable in that, being in the leading Hawksworth-designed coach, it not only hunted

Forty-three-year-old, Taunton-allocated, Collett-designed Mogul 6326 arrives into Dulverton with the 1603 Taunton to Barnstaple Junction. The locomotive was withdrawn that September and the 45-mile line across Exmoor closed in October 1966. Note the disused former Exe Valley line platforms on the right.

An extract from the local Women's Institute scrapbook. (Courtesy of Ansty WI Somerset Federation)

LAMENT

The single line winds its way up one of the steepest gradients in the country from Dulverton to Anstey. Under one bridge and then another, and the one line swells into two, between the two bare platforms. At night two Tilley lamps swing in the wind, casting weird shadows and quite failing to light the name ANSTEY, small and rusting on its post. The late traveller alights doubtfully at this lonely place, wondering if he has arrived anywhere. But there are some buildings— a signal box, ticket office and even a waiting room, and further down a goods yard and siding. For this was once a busy little station, lifeline to a large farming community and market. But now the line is doomed—the Beeching axe is poised to fall, and 1965 may well be its last year Already the staff has almost vanished—somewhere a boy may be found to sell you a ticket, but he also works the signals, receives the parcels—goods are already a thing of the past, so is the fire in the waiting room; last year's fag ends still litter under the seats. And the dear old puff-puff is superseded by a busy practical Diesel, which rolls in and out four times a day each way. The moss is growing on the platforms and as the Diesel grunts its way down the straight towards Barnstaple through the lovely intimate country of North Devon, there is a great sadness in the knowledge that soon this line, like so many others, will be empty and useless.

Contributed by A.B-C. to the Anstey volume of the Women's Institute Jubilee Scrapbook 1965.

Table 64

Taunton to Barnstaple and Ilfracombe

Weekdays — NOT on Saturdays 20 June to 5 September

Saturdays — 20 June to 5 September

				FSX	FSO									
TAUNTON ... ...d	7 45	11 25	13 15	16 20	17 55	19 50	21 10	6 10	7 00	..	8 30	11 25	12 55	13 27	16 03	..	17 45	21 10	..
MILVERTON .....d	7 57	11 37	13 28	16 32	18 07	..	20 02	21 22	..	..	7 12	..	8 41	11 37	13 08	..	13 41	16 16	..	17 57	21 22	..
WIVELISCOMBE .d	8 04	11 44	13 35	.	16 40	18 15	.	20 12	21 22	..	6 29	7 19	..	8 48	11 44	13 17	..	13 48	16 24	..	18 05	21 29	..
VENN CROSS .. ..d	8 15	11 55	13 46	..	16 51	18 26	.	20 23	21 40	..	..	7 30	..	8 59	11 55	..	..	.	16 35	..	18 16	21 40	..
MOREBATH HALT ...d	8 21	12 01	13 52	.	16 57	18 32	.	20 29	21 46	..	..	7 36	..	9 05	12 01	..	..	.	16 41	.	18 22	21 46	..
MOREBATH JUNCTION HALT. ... ...d	8 25	12 05	13 56	.	17 01	18 36	..	20 33	21 50	..	..	..	..	9 10	12 06	..	..	.	16 46	...	18 27	21 50	.
DULVERTON .....d	8 34	12 11	14 04	..	17 09	18 49	..	20 39	21 57	..	6 55	7 48	..	9 17	12 12	13 46	..	14 17	16 54	..	18 34	21 57	..
EAST ANSTEY .. ...d	8 43	12 20	14 13	.	17 18	18 58	.	20 48	22 06	..	..	7 58	..	9 30	12 21	14 00	..	14 26	17 03	..	18 43	22 06	..
YEO MILL HALT ......d	8 47	12 24	14 17	.	17 22	19 02	..	20 52	22 10	..	..	..	..	9 35	12 26	..	..	.	17 07	..	18 48	22 10	..
BISHOP'S NYMPTON & MOLLAND ......d	8 54	12 31	14 24	..	17 30	19 09	..	20 59	22 17	..	7 13	8 08	..	9 41	12 34	..	..	14 38	17 14	..	18 54	22 17	..
SOUTH MOLTON ...d	9 03	12 39	14 36	.	17 39	19 17	.	21 07	22 25	.	7 24	8 17	..	9 49	12 43	14 20	..	14 48	17 22	..	19 05	22 25	.
FILLEIGH. ....d	9 12	12 48	14 45	..	17 48	19 26	..	21 16	22 34	..	7 34	8 26	..	9 58	12 52	..	..	14 57	17 31	..	19 14	22 34	..
SWIMBRIDGE ...d	9 19	12 55	14 52	..	17 55	19 33	.	21 23	22 41	..	7 42	8 33	..	10 05	12 59	..	..	15 06	17 38	..	19 22	22 41	..
BARNSTAPLE JUNCTION... ...d	9 30	13 05	15 02	..	18 05	19 43	..	21 33	22 51	..	7 53	8 43	..	10 15	13 11	14 46	..	15 16	17 48	..	19 32	22 51	..
59 ILFRACOMBE .. .. a	10 15	13 47	15 51		19 00	20 53	..	.		.	8 45	9 37	..	11 15	14 41	15 33		16 13	19 02		21 02	.	..

Ilfracombe to Barnstaple and Taunton

Weekdays — NOT on Saturdays 20 June to 5 September

Saturdays — 20 June to 5 September

59 ILFRACOMBE ... d	..	6 48	8b20	12 15	..	15 20	.	17 57	..	7 55	9 25	10 12	.	11 05	12 20	14 55	..	17 57	18 50	..
BARNSTAPLE JUNCTION... d	6 40	.	8 35	10 40	14 10	..	16 10	..	18 50	6 48	.	8 43	10 15	11 05	..	11 58	13 11	15 50	..	18 37	19 34	..
SWIMBRIDGE .. ...d	6 49	.	8 44	10 49	14 20	.	16 20	.	19 00	7 02	..	..	11 19	..	..	16 03	..	18 48	19 45	..
FILLEIGH. ....d	6 57	.	8 52	10 57	14 28	.	16 27	.	19 07	7 10	..	..	11 27	..	..	16 09	..	18 56	19 53	..
SOUTH MOLTON ...d	7 08	.	9 04	11 08	14 40	.	16 40	.	19 17	7 21	..	9 12	10 41	11 38	..	12 26	13 43	16 17	..	19 04	20 02	..
BISHOP'S NYMPTON & MOLLAND .. ....d	7 16	.	9 12	11 16	14 49	.	16 47	.	19 26	7 30	..	..	11 47	..	..	13 47	16 29	..	19 12	20 11	..
YEO MILL HALT ......d	7 22	.	9 18	11 22	14 55	.	16 53	.	19 32	7 37	..	..	11 54	..	..	16 36	..	19 19	20 18	..
EAST ANSTEY .. ...d	7 28	.	9 24	11 28	15 00	.	17 00	.	19 38	7 42	..	..	12 00	..	..	14 00	16 41	..	19 24	20 23	..
DULVERTON ...d	7 38	.	9 34	11 38	15 10	.	17 09	.	19 46	7 52	..	9 40	11 08	12 10	..	12 51	14 10	16 52	..	19 35	20 34	..
MOREBATH JUNCTION HALT..... ...d	7 42	.	9 38	11 42	15 14	..	17 13	.	19 50	7 58	..	..	12 16	..	..	16 58	..	19 40	20 39	..
MOREBATH HALT ...d	7 46	.	9 43	11 46	15 18	..	17 17	.	19 54	8 01	..	..	12 20	..	..	17 02	..	19 43	20 43	..
VENN CROSS ... .d	7 54	.	9 51	11 54	15 26	..	17 25	.	20 02	8 09	..	..	12 28	..	..	14 28	17 10	..	19 52	20 51	..
WIVELISCOMBE .. ..d	8 06	.	10 01	12 06	15 36	.	17 35	.	20 12	8 19	..	10 09	11 33	12 39	..	13 16	14 39	17 20	..	20 02	21 01	..
MILVERTON ....d	8 12	..	10 07	12 12	15 42	..	17 41	.	20 18	8 25	..	..	12 45	..	..	14 45	17 26	..	20 08	21 08	..
TAUNTON .. ... a	8 25	.	10 20	12 25	15 55	..	17 55	.	20 30	8 37	..	10 28	11 50	13 00	..	13 35	14 57	17 38	..	20 20	21 20	..

Heavy figures indicate through carriages
For general notes see page 49

b Until 4 September dep. 8.48

Timetable extract of the Taunton to Barnstaple line.

at speeds over 20mph but had compartments with very upright ribbed seats. The resultant vibrations were like those felt by a flea on the tail of a friendly dog, carriage couplings and springs groaning in unison with the train's progress. Somehow this appalling rust bucket got us to Taunton on time, and unsurprisingly it was dispatched to the breaker's yard weeks later. After such a bone-shaking journey, the homebound Paddington train was sheer luxury as I was travelling over what was a novelty back then: continuous welded rail. Also conducive to the sleep-inducing environment was another factor not encountered over the past few hours – draught-free, warm accommodation.

The Great Western Steam Retreat

Although the WR timetable on p. 39 was dated until June 1965, a substantial September supplement declared all through services from Waterloo to the west of Exeter were to cease. The exceptions were the Brighton/Plymouth and the overnight newspaper 0110 ex Waterloo. The replacement, a semi-fast, two-hourly, Warship DL-worked service, was supplemented with stopping DMUs from Salisbury, some of which continued to Plymouth or Ilfracombe. Large groups of enthusiasts were seen on that final weekend and, having duly documented the proceedings, dispersed like vultures to witness the death throes of steam in other locations throughout the British Isles. However, there can be no other line in the whole country that saw more 'last steam' specials over such a lengthy period. More by luck than judgement, however, I was on the very last in November 1966.

The Exmouth Junction steam allocation of eighty-plus was massacred that September, with the lowest-mileage and 'fittest' Bulleids being dispatched to various SR sheds and leaving the bulk of the remaining allocation consisting of about three dozen GWR and Ivatt tanks.

Come January 1965 and all WTT (working timetable) services west of Exeter were turned over to DMUs – the irregular substitute service with substantial gaps throughout the day being designed for operating convenience rather than passenger suitability. March 1965 saw the Halwill to Torrington line close, followed seven months later by the Torrington to Barnstaple line. That summer saw numerous Salisbury-resourced steam substitutions for DL failures and, with the finite DL duties not being able to cater for the few remaining short-

WR withdrawals

The Western Region has announced that because of the economic crisis further cuts in train working costs are planned by speeding up the withdrawal of a number of uneconomic and lightly loaded passenger, parcels, and freight train services. Details of the passenger trains to be withdrawn are being discussed with the staff concerned, but the aim is to reduce local passenger train mileage by about a million train miles a year—a cut of about 6 per cent. It is proposed to achieve this mainly by timetable rationalisation but about one-fifth of the total will result from the closure of the Clevedon (Somerset) branch, the Taunton - Barnstaple line; the line between Okehampton and Bude and between Halwill Junction and Wadebridge; and by recasting the local stopping train services on the Bath to Weymouth route following the closure of nine stations. Ministerial consent for the withdrawal of services on these lines has already been given. Services proposed for rationalisation are those between Ealing and Greenford, Slough and Windsor; Reading and Newbury; Paddington and Oxford (suburban trains); Exeter, Barnstaple and Ilfracombe; Exeter and Kingswear; Plymouth local services; and on the Exmouth branch, also services between Swansea and West Wales, and intermediate services from Cardiff to Shrewsbury and Birmingham, via Hereford. It is hoped to implement all the proposals during October and November. Details will be announced locally. Among the results of the rationalisation measures will be the release of some diesel locomotives and multiple units for use elsewhere on BR.

Extract from Railway World magazine.

dated seasonal extras, through trains such as the sole Waterloo to Exmouth retained steam power over the former L&SWR route; I personally witnessed Unmodified 34063 *229 Squadron* on 31 July. On Sundays, the short-dated 0920 Salisbury to Exeter and 1936 trains were also frequently steam powered.

However, those lines that had escaped the 1965 axe didn't have long to wait. The October 1966 closures included both the Okehampton to Bude/ Wadebridge (abetted by the absence in the WR summer 1966 timetable!) and the Barnstaple to Taunton lines, followed by the Okehampton to Plymouth line in 1968. The Ilfracombe branch became a surprising post-Beeching casualty in October 1970, followed in 1972 by the Okehampton stub.*

---

* Following a lengthy public consultation, it is anticipated that Okehampton will be reconnected to the national rail network (with a branch of the Exeter/Barnstaple line at Crediton) in late 2021.

5

Varsity Line Visit

In between the two July 1964 outings to the Withered Arm, I squeezed in a visit to the cross-country Cambridge to Oxford line – for which closure notices had once again (!) been published.

This 77-mile route was known as the Varsity Line, the definition of the word varsity being a thing (in this case, a railway) that relates to universities. The route was opened in three separate stages: Bletchley to Bedford in 1846, Bletchley to Oxford in 1851 and eventually Bedford to Cambridge in 1862. With all companies involved being absorbed within London North Western Railway (LNWR), until 1922 they did not operate it as a through line and passengers had to change trains at Bletchley. Although BR attempted to close it, it was not included in the Beeching plans of 1963. However, with closure once again threatened, and eventually implemented in January 1968, I thought I had better, to quote the Beach Boys' current chart topper, 'get around' to travelling over the line before it was too late.

Having taken half a day of annual leave, I departed out of Liverpool Street on the D6722-worked 1236 for Cambridge, which conveniently connected into the second through train of the day, the 1412 Oxford departure. Locating a corner seat in a very warm DMU, Derby-built driving trailer M56227, I then endured a two-and-three-quarter-hour diesel-fumed journey on a blisteringly hot day. After crossing the flat Cambridgeshire

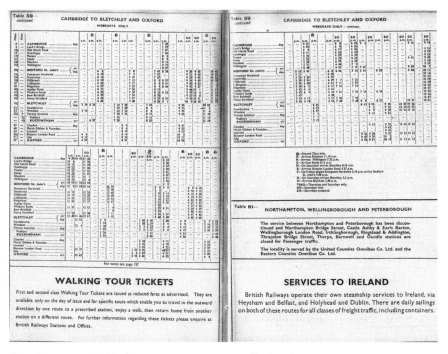

Timetable extract (LMR) of the east–west service on offer that day.

landscape and calling at the East Coast Main Line (ECML) junction station of Sandy, we paralleled the Great Ouse prior to stopping at the single-platformed Bedford St Johns, a station closed in 1984 upon the sole surviving section, now marketed as the Marston Vale services, from Bletchley being diverted into the Midland Main Line station at Bedford.

On we motored, passing the extensive London Brick Company works at Stewartby before our fifteenth stop at Bletchley. The scenery was now a little greener and after crossing under the GCR line at Calvert, we arrived at Bicester London Road. The line from here has been subsequently reopened to Oxford in 1987 and now is served via a chord coming in from the former GWML by regular services from Marylebone. Eventually we arrived into Oxford, not at the original Rewley Road station (1851–1951), but at the 1852-opened GWR station. Phew – what a sweat – but retrospectively, I was glad I made the effort. Over the years I had often forfeited visiting diesel-worked lines doomed for closure in order to prioritise costs in favour of known steam-operated lines. That day's outing cost a pricey 12s 11d (65p).

It was a hot and sunny Tuesday, 14 July 1964, and having just arrived into Oxford on a three-hour DMU journey from Cambridge, I leaned out of the window for some air – capturing Swindon's Modified Hall 7916 *Mobberley Hall* (a seventeenth-century country house in Cheshire) readying herself for departure with the 1725 local for Banbury.

In woeful external condition, Collett 4-6 0 6916 *Misterton Hall* pollutes the sky as she accelerates her southbound freight through Oxford. She was one of 258 built at Swindon between 1928 and 1943, this particular Banbury-allocated example being withdrawn in July 1965.

One of the last locomotives of pre-nationalisation designs to be built, this 1950 Robert Stephenson & Hawthorn Ltd-built 0-6-0PT 9411 rests in the evening sun at Paddington. Old Oak Common-allocated, she was withdrawn upon 81A's closure to steam the following spring.

This was my first visit to Oxford (thus justifying this chapter's inclusion in the book) and during the time I spent there I viewed and photographed a succession of steam locomotives on freights, together with a Modified Hall on a Banbury stopper. Indeed, as I was to witness in my later outings there, examples of steam classes from all regions were to be seen at this railway crossroads. Wow, I thought, I must return here in order to search out what steam-hauled passenger trains are on offer. Regrettably, with the fast-changing steam scene throughout Britain luring me elsewhere, I failed to carry out my promise for a further eight months.

6

A Double Whammy

We now move on to August 1964 and on Friday the 7th, I took half a day of leave to sample another run with GWR power over the cross-country route from Redhill via Guildford to Reading. Plans had been announced that this line was to be worked by hybrid 'Tadpole' DEMUs from January 1965 and, having read reports that it was almost entirely steam operated, I made a beeline for the WR-resourced 1331 arrival into Reading (Southern), on this occasion reaping a run with Reading's 7829 *Ramsbury Manor*.

The Reading station I arrived into that day was opened in 1849 by the South Eastern Railway, which operated its trains through from London and Surrey. Seven years later, courtesy of a line from Ascot joining the route at Wokingham, regular services operated from London Waterloo by the L&SWR were added to the mix. Renamed Reading (South) in 1949 in order to distinguish it from the adjacent WR station, a further renaming to Reading Southern took place in 1961 before the four-platform terminus closed to passengers in September 1965 upon the provision of an electrified bay (4A) at the main station.

Making my way the short distance to the 1906-opened Reading West station in order to travel on the recently diverted (from the S&D) Pines Express, another brief paragraph of historical data is due.

A Double Whammy

Friday, 7 August 1964 and, having just worked into Reading (Southern), home-allocated 7829 *Ramsbury Manor* (a privately owned Wiltshire house that when sold at £650,000 in 1966 was listed in *The Guinness Book of Records* as the most expensive in Britain) waits for the signals to enable her to return, via the General station, to the shed. A Carmarthen-allocated locomotive for many years, her last shed was to be Gloucester. She was one of the two final class survivors, being withdrawn at the very end of WR steam.

This is Reading West, a wayside halt where a considerable amount of my time was spent awaiting late-running inter-regional trains. Here Bournemouth's 34040 *Crewkerne* snakes her way in off the spur with the southbound Pines Express. This all-year through service from Manchester to Bournemouth had been deliberately diverted away from S&D metals as a prelude to the usual 'closure by stealth' method often implemented by the authorities during the 1960s.

Constructed by GWR, the line between Reading and Basingstoke was opened in 1848 but, being broad gauge, it initially had its own terminal at Basingstoke to the north of the L&SWR station. Although GWR was ordered to convert to standard gauge in 1854 (to be implemented by February 1856), it was not until December 1856 that the line was (re) opened as a mixed-gauge railway and, with typical obstinacy, GWR kept its station open at Basingstoke until 1932. By the construction of a forty-five-chain link from the Didcot line, through trains from the north of England to the south coast avoided the necessity to reverse at the General station. The chapter title 'Double Whammy' can now be elucidated upon because, in my naivety, I had fully expected the Pines Express to be worked by GWR power – only to find it disappointingly worked by filthy Bournemouth-allocated West Country 34040 *Crewkerne*.

Catching it to Basingstoke was a deliberately designed strategy to be in position for the 1512 stopper to Woking, which had been reported in *Railway World* as being S15 worked. What did I get? Guildford's N Mogul 31858! There were many highs and lows during those years – with disappointments, such as the double whammy collected that day, quickly forgotten when either successes or unusual locomotives were captured days or weeks later.

7

That Middle
England Jaunt

Also in August 1964, I enjoyed a week-long stay with my father's relations in Leicester. With various local outings, together with excursions to Nottingham and Peterborough, it was on Thursday the 27th that I made my first visit to Birmingham. Travelling via Nuneaton into Birmingham New Street and equipped with Lt Aiden L.F. Fuller's well-thumbed *British Locomotive Shed Directory*, I walked the short distance to Snow Hill, passing shops blaring out either Manfred Mann's 'Do Wah Diddy Diddy' or the Honeycombs' 'Have I the Right', both of which were vying for the top spot that month.

Although the GWML from Paddington to Chester had, in the 1963 boundary changes, been awarded to the LMR north of Aynho Junction, Birmingham's Snow Hill station was still very much steeped in its former owner's paraphernalia and atmosphere. Opened in 1852 by GWR, it was initially named just Birmingham, then Great Charles Street, then Livery Street before Snow Hill was settled upon in 1858. Starting off as a nondescript wooden construction, a major revamp in 1871 equipped the station with a glazed overall roof, two through roads and two north end bays. The prohibitive cost of widening the 635yd-long tunnel at the south end of the station resulted in Moor Street being born in 1909 to deal with the burgeoning suburban traffic from Leamington and Stratford. Further expansion a few years later to compete with London, Midland and Scottish Railway's (LMS) New

Rebuilt by Collett from the frames of Churchward's 2-6-2T 5115, Prairie 8109 was the last survivor of the once ten-strong class. Passing through the cavernous Birmingham Snow Hill on Wednesday, 26 August 1964, she was withdrawn at Tyseley in June 1965.

Street services saw the trackwork north of the site quadrupled, together with a major upgrade to the station I was witnessing that August visit.

You could see at one time not so long ago it was the jewel in the crown of GWR expresses to/from all parts of the country. With its wide (once carpeted as a marketing gimmick by Cyril Lord) staircases leading to the two major platforms, this once great cathedral of steam was now suffering from an overdose of fumes from DMUs ticking over. It was a hot and sunny day, and the somewhat claustrophobic effect created by the tunnels at both ends made it a sweaty, uncomfortable visit; the aforementioned odours didn't help. The one resident Pannier tank busying itself shunting parcel vans at the north end of the station (reputedly the best position for trainspotting and photographing), together with an elderly tank passing through light engine, just about made the stay marginally more tolerable.

Now, however, the station had an eerie, underused atmosphere about it. The roof and walls were blackened with soot and smoke from many years of steam locomotives idling there to allow the thousands of passengers to alight or join their trains. Of course, at the time I was naively unaware of the Kings and Castles that once dominated those trains, they having been long replaced by diesel traction. Compared with New Street, being rebuilt in

Still retaining all her number and nameplates (an increasing rarity for all former GWR locomotives by that date), home-allocated 6922 *Burton Hall* performs empty coaching stock (ECS) duties at Shrewsbury.

Just days away from her withdrawal, yellow-striped (denoting her being banned from working south of Crewe under the overhead line equipment) Jubilee 4-6-0 5XP 45577 *Bengal* stands at Shrewsbury awaiting a signal for the shed. She was one of 6D's three surviving Jubilees.

connection with the West Coast Main Line (WCML) electrification, Snow Hill was a throwback to an earlier time. Indeed, in March 1967, upon cessation of the Paddington to Birkenhead expresses, it went into a steep decline. With the majority of suburban services to the south diverted to Moor Street and trains from Stourbridge and Shrewsbury diverted to New Street, just a handful of local stopping services to Wolverhampton remained. These were inevitably withdrawn in 1972 and the station closed; the site was razed to the ground some five years later. With the saying 'what goes round comes around' in mind, in 1987, with the roads gridlocked, it was resurrected, albeit initially as a two-platform concrete structure, as part of the Midland Metro – the full circle being completed some years later when the Marylebone to Kidderminster services were implemented.

The myriad lines in the Birmingham and Wolverhampton area were, to a 17-year-old, of bewildering complexity. Having negotiated them on my Brush-hauled, Birkenhead-bound train, I was glad to make the open countryside of Shropshire, destined for its county town of Shrewsbury.

For sure, upon arrival there it seemed that a fair percentage of the passenger trains (other than the ones I was travelling on!) were steam operated and, although vowing to return one day to sample them, it was to be a further year before I was to return. As a far from experienced traveller, I feared straying from my planned circular itinerary and left Shrewsbury with yet another Type-4-hauled train for the 32-mile journey onward to Crewe.

Growing confidence over the following few years would mean that I would board a train if it was being worked by a 'never travelled with before locomotive' – the addictive rationale of a seasoned haulage basher – wherever it was destined or even not knowing its next stop.

8

Pastures New

It was now September 1964 and no fewer than three popular TV series commenced broadcasting that month: *The Addams Family*, *The Munsters* and *The Man from U.N.C.L.E.*, while the first issue of *The Sun* (replacing *The Daily Herald*) was published. Musically, the Kinks, who went on to attain celebrity status among steam enthusiasts with their 'Waterloo Sunset' release commemorating the end of SR steam in July 1967, held the number one spot for two weeks with 'You Really Got Me'.

As for my railway travels, two doomed lines – the truncated remnant of the 24-mile line from Worcester to Leominster (as far as Bromyard) and the Hereford to Gloucester via Ross-on-Wye – had caught my eye. And on Thursday the 3rd, I headed to a part of the country I had yet to visit.

It was totally new territory for me and, after Hymek diesel-hydraulic D7061 had powered me out of Paddington on the 0915 service to Worcester, rather disappointingly (in my naivety I was expecting at least a two-vehicle train headed by perhaps a GWR Pannier!), after an hour's wait, the 1305 Worcester Shrub Hill to Bromyard was formed of a run-of-the-mill DMU.*

---

* DMU stood for diesel multiple unit – better known to us teenagers as 'bog-carts'. A despised hybrid, these trains were neither carriage nor loco-hauled and many enthusiasts (myself included) did not even bother to take down their numbers. DMUs offered a no-frills, bog-standard travel – hence their name. They had dirty windows, which rattled in their frames, seats that vibrated obscenely if not weighed down and engine fumes that often seeped inside the carriage to give you a thick head. The only plus was being able to sit behind the driver (if he hadn't pulled the blinds down) to see the way ahead just as he did and watch his practised hand poised on the controls. (Nicholas Whittaker, *Platform Souls*.)

Long-term Worcester resident 1942-built 4F 4613 performs ECS duties at the Shrub Hill station on Thursday, 3 September 1964. A colossal 863 of these Collett-designed Pannier tanks were built over a twenty-year period from 1929 to 1949.

Unwanted and unloved 4-6-0 5P 5096 *Bridgewater Castle*, having been withdrawn three months earlier, stands outside Worcester shed. Taken from the window of my Bromyard-bound train, if I read my scrawled notes accurately, she was strangely adorned in a puerile green undercoat. She had been reallocated between fourteen sheds during her twenty-five-year life.

Pastures New

The Royal Assent for construction of the 27½-mile railway between Worcester and Leominster was granted in 1861 but it took an astonishing thirty-six years before the line was completed due to contractual difficulties, bankruptcy and liquidation problems. Although the Bromyard races and seasonal hop-picker traffic swelled the coffers, the line was always only ever lightly used and the section west of Bromyard was closed to passenger traffic in 1952. The final passenger service that ventured all the way through to Leominster was a 1958 rail tour. A journey along the remaining 14½-mile stub took a mere twenty-nine minutes and during the twelve-minute turnaround I unwisely didn't take any photographs. A return daylight trip was only possible on a Thursday or Saturday; look at the appalling timetable on offer (p. 66). The station and line closed the following Monday.

Having returned to Worcester, I alighted at the Foregate Street station in order to head west to Hereford (née Hereford Barrs Court for the first forty years of its life) in yet another vexatious DMU.

At last matters improved with both Hereford and the 22-mile line via Ross-on-Wye awash with steam. Opened in 1855 and converted from broad gauge fourteen years later,* the northern section of this delightfully rural and scenic railway required extensive engineering to cross the meanders of the River Wye on no fewer than four occasions with either embankments or short tunnels necessary at each one. Climbing on through Ross-on-Wye, the line peaked in the 771yd Lea Line tunnel before descending through the Forest of Dean. However, scenery doesn't make for profitability: in two months' time the line was to lose its passenger trains, with freight over the southern section lingering on into the following year. The power for my 1630 Hereford to Gloucester was 86C's Collett 5101-class Prairie 4161, which, after its home depot closed just eight weeks later, went on to survive until the end of November 1965 at Worcester.

---

* This was an early replacement of broad gauge to narrow gauge, this line being selected as being a quieter track to gain experience. For two weeks from Monday, 16 August 1869, horse buses replaced train services. Two days earlier, three special trains conveying 400 workers arrived from South Wales, bringing with them a week's supply of food. Sleeping accommodation was provided by forty broad-gauge wagons, which had been specially cleaned, whitewashed and lined with straw to create bedding for the workers. Those in charge had the luxury of a first-class carriage. Work along the 22-mile line progressed so well that all was completed by the Friday, allowing the workers to collect their wages earlier than anticipated and returning home to no doubt celebrate with a few drinks.

Table 51 — Weekdays only

Gloucester, Ross-on-Wye and Hereford

												C	D SO					SX	SO	SX	SO
GLOUCESTER CENTRAL.	d	7 00	9 48	12 15			14(25	14(47	16 00	17 55	19 15		21 40	21 55							
OAKLE STREET	d	7 09	9 56	12 23			14(33	14(55	16 08	18 04	e		21 48	22 03							
GRANGE COURT A	d	7 14	10 01	12 28			14(38	15(05	16 13	18 09	19 27	19 27	21 52	22 07							
BLAISDON HALT.	d	7 18	10 05	12 32			14(42	15(04	16 17	18 13	c	19 31	21 57	22 12							
LONGHOPE.	d	7 25	10 11	12 38			14(48	15(10	16 23	18 18	e	19 36	22 03	22 18							
MITCHELDEAN ROAD	d	7s37	10 18	12 45			14(55	15(17	16 30	18 25	e	19 43	22 10	22 25							
WESTON-UNDER-PENYARD HALT	d	7 42	10 23	12 50			15(00	15(22	16 35	18 29	f	19 48	22 16	22 32							
ROSS-ON-WYE	a	7 45	10 27	12 54			15(04	15(26	16 40	18 32	19 51	19 52	22 20	22 35							
	d	8 00	10 30	12 56			15(06	16(29	17 00	18 38	19 52	19 53	22 21	22 36							
FAWLEY	d	8 11	10 42	13 07			15(17	15(40	17 10	18 48	e	20 04	22 29	22 44							
BALLINGHAM	d	8 14	10 46	13 10			15(20	15(43	17 13	18 51	e	20 07		22 48							
HOLME LACY	d	8 21	10 54	13 17			15(27	15(50	17 20	18 58	e	20 14	22 30	22 51							
HEREFORD	a	8 30	11 04	13 29			15(39	16(00	17 30	19 09	20 21	20 24	22 47	23 05							

Hereford, Ross-on-Wye and Gloucester

						D SO	C	C		D SO				SX	SO	
HEREFORD	d	6 55	7 30	10 25	13(25	13(40	14(38	14(38	16 30	18 02	21 15	21 42				
HOLME LACY	d	7 04	7 39	10 33	13(33	13(49	14(46	14(46	16 39	18 11	21 23	21 51				
BALLINGHAM	d		7 45	10 39	13(39	13(55	14(52	14(52	16 45	18 17	21 30	21 57				
FAWLEY	d	7 12	7 48	10 42	13(42	13(58	14(55	14(55	16 48	18 20	21 33	22 00				
ROSS-ON-WYE	a	7 20	7 57	10 52	13(51	14(07	15(04	15(05	16 57	18 30	21 42	22 08				
	d	7 24	8 07	10 53	13(52	14(10	15(06	15(06	17 00	18 34	21 43	22 12				
WESTON-UNDER-PENYARD HALT	d	7 29	8 12	10 58	13(56	14(15	15(11	15(11	17 05	18 39	j	22 18				
MITCHELDEAN ROAD	d	7 35	8 18	11 04	14(02	14(21	15(18	15(18	17 11	18 45	21 53	22 26				
LONGHOPE.	d	7 41	8 24	11 10	14(08	14(27	15(24	15(24	17b22	18 51	22b04	22 34				
BLAISDON HALT.	d	7 46	8 29	11 15	14(13	14(32	15(28	15(28	17 27	18 56	e					
GRANGE COURT A	d	7 50	8 33	11 19	14(17	14(36	15(33	15(38	17 31	19 01	22 14	22 45				
OAKLE STREET	d	7 54	8 37	11 23	14(21	14(40	15(42	15(47	17 35	19 05	22 18					
GLOUCESTER CENTRAL.	a	8 02	8 46	11 33	14(30	14(49	15(53	15(58	17 47	19 15	22 27	22 58				

Heavy figures indicate through carriages. For general notes see page 49

A For other trains between Gloucester and Grange Court see Table 80
C Not on Saturdays 20 June to 5 September
D Until 5 September

b Arr. 5 minutes earlier
c Stops to set down on notice being given to the guard at Gloucester
e Stops to set down on notice being given to the guard at the previous stopping station

f Stops to set down passengers from beyond Gloucester on notice being given to the guard at Gloucester
j Stops to set down on notice being given to the guard at Hereford or Ross-on-Wye

Table 52 — Weekdays only
Worcester and Bromyard (Second Class only)

			SO		ThSO					SO	
WORCESTER {SHRUB HILL	d	10 20		13 05		16 10		17 45	22 15	
{FOREGATE ST	d	10 23		13 10		16 23		17 51	22 19	
HENWICK B	d	10 26				16 26		17 54		
RUSHWICK HALT B	d							17 57		
LEIGH COURT	d	10 37		13 20		16 35		18 06	22 33	
KNIGHTWICK	d	10 44		13 28		16 42		18 13	22 40	
SUCKLEY	d	10 49		13 33		16 47		18 18	22 45	
BROMYARD	a	11 00		13 44		16 58		18 29	22 56	

Bromyard and Worcester (Second Class only)

			SO		ThSO					SO	
BROMYARD	d	7 40	11 10	13 56	17 15	18 50	23 05	
SUCKLEY	d	7 54	11 21	14 07	17 27	19 01	23 16	
KNIGHTWICK	d	7 58	11 26	14 12	17 33	19 05	23 22	
LEIGH COURT	d	8 05	11 32	14 19	17 39	19 13	23 30	
RUSHWICK HALT B	d	8 13						
HENWICK B	d	8 17	11 43	14 30	17 50			
WORCESTER {FOREGATE ST	a	8 19	11 45	14 32	17 52	19 24	23 45	
{SHRUB HILL	a	8 25	11 49	14 38	17 56	19 29		

Heavy figures indicate through carriages. For general notes see page 49

B For other trains between Worcester, Henwick and Rushwick Halt see Table 41

282

Whilst waiting for my Hereford-bound train, 21-year-old 6956 *Mottram Hall* (a 175-built Cheshire house now in use as a luxury hotel) passed through Worcester Foregate Street with an eastbound freight. This then-Gloucester-allocated locomotive was withdrawn at Oxford at the end of 1965.

At Rotherwas Junction (south of Hereford) Gloucester-allocated 4-6-0 Modified Hall 6985 *Parwick Hall* is passed – this 16-year-old being withdrawn later that month.

Hereford-allocated 0-6-0 2287 at Ross-on-Wye crosses us with a northbound freight. Heavy parcels traffic as seen on the platform failed to save the line, closure being just eight weeks away.

Gloucester Central was another revelation. Unable to encompass a trip on the Chalford 'auto' (also ceasing that November), at least I took a photograph of one – together with a couple of the seemingly endless procession of steam-operated freights passing through. They were just a foretaste of what was to be the biggest surprise of the outing.

Notes have long been lost from that day, but having selected the 1815 departure for Swindon as a means to return to London, I cannot recall what brought the train from its starting station of Cheltenham. However, I well remember the exhilaration at what took over: one of the last dozen or so Castles, 7005 *Sir Edward Elgar*, which was to take me the 36¾ miles to Swindon via the Golden Valley that evening en route home to London. Referring back to the final sentence of Chapter 6 – this was one of the highs.

Despite widespread dieselisation elsewhere, this route was allowed to continue with steam, with no attempt to save money or offer a better service. My train, the 1630 Hereford to Gloucester Central, pauses at Longhope, where I took the opportunity of a lengthy station stop awaiting an opposite way working over the single line to alight onto the platform to photograph the train loco, Collett 5101 class 2-6-2T 4161.

One of the famed Golden Valley auto trains, the 1715 from Chalford, is seen after arrival into Gloucester Central with Horton Road's 0-4-2T 1444. Beeching cuts led to the service ceasing in November that year.

Passing through Gloucester Central is home-allocated Modified Hall 7926 *Willey Hall* with an eastbound freight.

The penultimate standard-gauge steam locomotive built in Britain, Cardiff East Docks-allocated BR 9F 2-10-0 92219, constructed at Swindon in January 1960, eases her way through Gloucester Central on a westbound freight. She was withdrawn in September 1965 after just five years and six months of service.

The 1745 Cheltenham St James to Swindon was, upon reversal at Gloucester Central, worked for the 36¾ miles to Swindon by Castle 4-6-0 7005 *Sir Edward Elgar*. With, by that time, just a handful of Castles remaining, I considered this quite a catch; indeed, she was withdrawn four days later. Originally christened *Lamphey Castle*, this lifelong Worcester resident was renamed in 1957 to honour the 100th anniversary of the Worcester-born composer's birth.

9

The Cambrian Crusade

Moving on to November 1964 and, spurred on by the news of widespread line closures throughout North Wales, I embarked on a twenty-four-hour, 600-mile 'hit' to travel over them before it was too late.

Making the headlines in politics that autumn was Harold Wilson's Labour Party, which secured a narrow victory of just four seats to end the thirteen-year tenure of Conservative rule – their first act being the abolition of the death penalty. As an aside, one of their manifesto promises, to halt the large-scale railway closures, was reneged upon! Musically, the three chart toppers that month were Sandie Shaw's '(There's) Always Something there to Remind Me', Roy Orbison's 'Oh, Pretty Woman' and the Supremes' 'Baby Love'. And who can forget the soap *Crossroads*, with its 'wooden' acting and Benny and Miss Diane, the first episode of which was broadcast in November.

Having suffered a disrupted journey from Kent, resulting from an unofficial go-slow by ASLEF union drivers, I met up with my travelling companions, Keith and Dave, at the building site that was allegedly Euston station (it was undergoing rebuilding in connection with the WCML electrification) in time to catch the 2235 departure for Crewe.

Our connection into Wales, the 0219 Crewe to Holyhead 'Mails', only recently reported by *Railway World* as a Britannia Pacific duty, was disappointingly worked that morning by 'Long Pong' D227 *Parthia*. The Isle of Anglesey section of the route was not coloured in on my BR map (as I didn't

Bulleid Light Pacific 34027 *Taw Valley* is at Holyhead on Tuesday, 31 July 1990 with the 14.56 North Wales Coast Express departure for Crewe. At Llandudno Junction, while the Pacific was taking water, Brush Type-2 DL 31102 hauled the train to Llandudno where, having slaked her thirst, *Taw Valley* came down the branch to work it onto Crewe.

traverse it with steam) until it was travelled over on the North Wales Coast Express in 1990 with SR Pacific 34027 *Taw Valley* while attending a railway training course at Webb College, Crewe. Although the objective of the trip was to travel over a great many lines proposed for closure, the dieselisation of this expected steam working certainly put the dampers on it and we sat/slept in silence during the two-and-a half-hour journey. As time went by, with an increasing amount of such substitutions, I learnt to 'get over' the associated feelings of disillusionment and focus on the positives when they occurred.

The Grand Junction Railway had opened the line we were travelling over in 1840 between Crewe and Chester, the remaining 84 miles to Holyhead eventually being completed by the Chester & Holyhead Railway Company in 1850 – the all-conquering LNWR absorbing it a few years later. So important was the passenger, mail and freight traffic between Britain and Ireland that the world's first water troughs were installed at Mochdre (Colwyn Bay) to enable steam locomotives to collect water without stopping. Certain sections east of Llandudno Junction were subsequently quadrupled to increase line capacity but have since reverted to two tracks.

Having arrived into Bangor sixteen minutes late at 0451, it gave us just enough time to bunk the shed, conveniently adjacent to the station, with the notes of the occupants having long been lost, prior to catching the 0520 Pwllheli departure behind Mickey tank 41234. Bangor shed, coded 6H, had a small allocation of Ivatt 2MTs for pilot duties, local passenger and freight work, which included workings to and from Amlwch and the Llyn seaside resort of Pwllheli.

The line we were about to travel over that morning was opened from a junction just short of the Menai suspension bridge as far as Caernarfon in 1852 but it was a further fifteen years before it was extended to meet Cambrian Railways at the purpose-built Afon Wen station. Once frequented by through trains from Manchester conveying thousands of holidaymakers to the Butlin's holiday camp at Penychain,* increasing car ownership and more direct coach travel had decimated train usage. Of all the lines traversed that day this was the most 'urgent', the section south of Caernarfon, originally proposed for closure in March 1964, succumbing just three weeks later. Even the huge crowds that attended the Prince of Wales' investiture at Caernarfon Castle in July 1969 failed to save the truncated stub and it closed the following January. Two sections of the line are now in use as a cycleway, another as part of the A487 road, while between Caernarfon and Dinas the Welsh Highland Heritage Railway has purloined it.

This 0520 departure was, as a result of heavy mail and paper traffic, the only non-DMU service each day, and it had no balancing working – at least, none advertised. This scenario was to become common during the latter years of steam, thus turning us haulage bashers into nocturnal travellers, solely to catch steam workings over routes that were otherwise diesel worked.

Having travelled through the inky darkness of a November morning, dawn was just breaking upon our arrival into Pwllheli at 0655, some 32 miles later. The station we arrived into that morning was of 1909 vintage, being constructed closer to the town than the original. Dave, being a bus fanatic,

---

* Penychain station was opened in 1933 to serve the nearby Butlin's holiday camp, which was under construction at the time. With the outbreak of the Second World War, the camp was requisitioned as a naval training base, returning to civilian use in 1947. The station was connected to the camp on summer Saturdays by a 'Puffing Billy' road train. Although much reduced in size, and now just a single platform, the station is still open (with an annual footfall of 4,000) and serves a Haven Holiday and caravan park on the former Butlin's site.

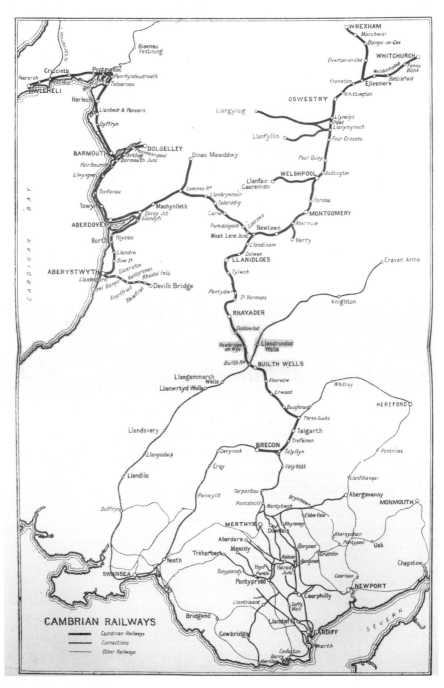

Map of Cambrian Railways system.

The Cambrian Railways coat of arms. Although the majority of its route miles lay in Wales, the headquarters of the company was in Oswestry in England.

insisted on a visit to the Crosville garage, and so a wander around the town before most inhabitants were stirring was the order of the day.

We were now on the territory of the former Cambrian Railways and so a potted history of the company is due. At its peak, Cambrian Railways (1864–1922) owned 230 miles of track over a large area of mid-Wales. Formed from an amalgamation of four smaller railways, a further eight, including the Vale of Rheidol and Welshpool & Llanfair narrow-gauge systems, were absorbed over the years to 1913. With its headquarters in the English town of Oswestry, employment at the accompanying works and locomotive depot resulted in the town's population doubling over the forty years from 1861, and at the 1923 grouping they handed over ninety-four standard-gauge and five narrow-gauge locomotives to the GWR.

At Pwllheli, our steed for the 0735 all-stations departure for Barmouth was BR 3MT 82020. The Machynlleth-allocated engine and a dozen of her sisters had been sent there to replace the Churchward 45xx and 4575 classes after the trials of a batch of Stanier 3MT 2-6-2Ts were deemed a failure. No. 82020 had herself arrived at 6F (née 89C) from Shrewsbury in March 1960 and would be forwarded to Nine Elms, still retaining her WR green livery, on 19 April 1965.

Opened throughout in 1867, this line, a masterpiece of railway engineering, took five years of sweat, toil and endurance to build, with the workers having to endure a constant battle against raging seas, crumbling cliff edges and adverse weather conditions. And what a wonderfully scenic line it is, with rocky peninsulas and shimmering sea on one side and Wales' highest mountain, Snowdon, on the other. Not all of this vista was available to us that day, with early morning mist shrouding the line, the leisurely pace being enhanced by frequent operation of the locomotive's whistle to disperse the myriad seagulls resting on the track ahead. After Porthmadog, where the Ffestiniog and Welsh Highland narrow-gauge railways can be accessed, we

The Cambrian Crusade

Six of Machynlleth's BR 3MTs were dispatched to Nine Elms during spring 1965: 82005/6/20/1/32/3. With 82032 having failed en route at Willesden and all the others having been withdrawn (because of their poor maintenance status) by the end of 1965, the sole survivor, green-liveried 82006, is seen at Clapham Yard in February 1966.

called at Minffordd, where nowadays fans of the cult 1960s TV series *The Prisoner* can alight to visit Portmeirion, perhaps to investigate what Number Six was escaping from!

After traversing the twenty-one-pier Victorian Pont Briwet viaduct over the River Dwyryd (subsequently rebuilt in 2014), we passed Harlech Castle, the three of us attempting to render our own version of the famous regimental march 'Men of Harlech' and unsurprisingly being frowned upon by those within hearing distance! This coast-clinging line, although a frequent casualty of storm-force winds and tidal surges from Cardigan Bay, has survived several attempts at closure. Although never covering its operating costs, it appears to have a secure future in that it is a vital lifeline, keeping the sparsely populated communities together. It is also heavily used by schoolchildren, fortunately absent this day as it was a Saturday!

After our arrival at Barmouth, the greed for more steam haulage kicked in, the service we originally planned to catch being the 1020 to Chester General via Bala Junction and Llangollen. However, rather than wait at the rain-sodden Barmouth station, we caught the Pwllheli portion of

Table 85—
continued

PWLLHELI, BARMOUTH and MACHYNLLETH

Miles			WEEKDAYS																	
			am		am	am		am	am	SX am		am		am			am	am		am
					TC Aberystwyth dep 7 35 am to Shrewsbury	TC to Wrexham General arr 9 45 am; Via Ruabon (Table 83)			TC Pwllheli to Bangor arr 7 44 am					TC Pwllheli to Bangor arr 9 14 am		CAMBRIAN COAST EXPRESS. TC Pwllheli to London Paddington. RB Shrewsbury to Paddington				TC to Chester General arr 1 21 pm; Via Ruabon (Table 83)
0	**Pwllheli**	dep	5 50						6 30			7 35		7 45			8 20			
2	Abererch Halt		dd									7 40								
3½	Penychain **D**		dd									7 43		7 52			8A28			
4¾	**Afon Wen**	arr	6 1						6 40			7 47		7 56						
		dep	6 8					6 35				7 55								
8	Criccieth		6 15					6 41				8 1					8 37			
9½	Black Rock Halt																			
13	**Portmadoc**	arr	6 23					6 50				8 10					8 46			
		dep	6 25					6 55				8 13					8 48			
15¼	Minffordd							7 0				8 18								
16¼	Penrhyndeudraeth		6 32					7 4				8 23					8 56			
17	Llandecwyn Halt							7 7				8 26								
18¼	Talsarnau		6 37					7 12				8 31								
19¼	Tygwyn Halt							7 15				8 34								
21¾	**Harlech**		6 43					7 20				8 42					9 7			
23¾	Llandanwg Halt							7 25				8 47								
24¼	Llanbedr and Pensarn		6 50					7 28				8 50								
25	Talwrn Bach Halt		6 52					7 30				8 52								
27¼	Dyffryn-Ardudwy		6 58					7 36				8 57								
28½	Talybont Halt							7 40				9 1								
30¾	Llanaber Halt		dd					7 44				9 6								
32½	**Barmouth**	arr	7 7					7 48				9 12					9 26			
		dep	7 11			7 18				8 15							9 31	9 45		10 20
33¾	Morfa Mawddach	arr	7 16			7 23				8 20							9 36	9 50		10 25
42¾	83**Dolgellau**	arr				7 43												1015		10 46
		dep															8H30			
—	Morfa Mawddach	dep	7 17							8 20							9 37			
35	Fairbourne		7 20							8 23							9 39			
37¾	Llwyngwril		7 30							8 31							9 48			
39½	Llangelynin Halt									8 35							dd			
41¾	Tonfanau		7 38							8 40							9 56			
44¼	Towyn		7 44							8 45							10 1			
47¾	**Aberdovey**		7 52														10 8			
48¾	Penhelig Halt		7 55														10 11			
50½	Abertafol Halt																dd			
52¼	Gogarth Halt																dd			
53¾	Dovey Junction	arr	8 6														10 28			
70¼	86Aberystwyth	arr	8 54														11 42			
—	Dovey Junction	dep			8 15												10 36			
57¾	**Machynlleth**	arr			8 22												10 43			
—	83Ruabon	arr				9 35														12 41
—	86Shrewsbury	„			10 35	10 39											12 47			1 35
—	86Birmingham (Snow H.)	„			11 55	11 55											1 55			2 55
—	86**London** (Paddington)	„			2 5	2 5											4 0			5 0

For Notes and Continuation of Trains, see pages 265 and 266

the Cambrian Coast Express across the Barmouth railway viaduct,* any road traveller having to take a 14-mile detour to Morfa Mawddach (née Barmouth Junction) with BR 4MT 80104. We then changed onto what proved to be the first of five runs with different Ivatt 2-6-0s that day – this one was 46521 on a local train to Dolgellau (the English spelling of Dolgelly was used by the railway until September 1960) and only then did we catch the 1020 train from Barmouth.

What followed then was a splendid 45-mile ride through mid-Wales. Opened throughout in 1868 by GWR, principally as a holiday route to the Cardigan coast resorts from the Wirral, in comparison to the coast-hugging line we had just travelled over, this rugged gradient-strewn line was the opposite. We slowly carved our way uphill through thickly forested areas of stunning autumnal magnificence, the hillsides above them swathed in the now amber-coloured dying fern. We then began climbing higher into the scenic Berwyn Hills, now able to view the snow-capped Aran Fawddwy in the distance.

After passing by the lengthy Bala Lake, we then paralleled the River Dee. The line weaved through deep gorges, past rushing rivers and, being November, nature was at her colourful best with the stark, forbidding outlines of the mountains contrasting wonderfully with the multi-coloured-leaved trees at the lower level. All this behind Croes Newydd's BR 4MT 75023, which was well up to the task with her sharp exhaust echoing off the trees and hillsides en route.

This line was proposed for closure later that month but this was delayed while adequate bus provision was supplied. However, severe flooding in the second week of December resulted in the section between Llangollen and Bala Junction being abandoned, the entire route ceasing operation on 18 January 1965. Two sections live on: the Bala Lake Railway operates a narrow-gauge line (over the former track bed) alongside the lake itself, while the Llangollen Railway operates trains from there to Corwen.

---

* The 764yd Barmouth railway viaduct over the River Mawddach, available to foot passengers and cyclists, was under threat of closure in the 1980s upon the discovery that the wooden piers had been eaten away by marine woodworm. After a six-month shutdown for repair work, although reopened for trains, loco-hauled services, i.e. freight and excursions from all parts of the country, were banned, seriously affecting the local economy, until further remedial work was carried out in 2005.

Forty-two years later and BR Standard 4MT 80104 happily lives on at the Swanage Railway.

Having spent the majority of her short, thirteen-year life at Brecon, Ivatt Mogul 46521 arrived at her final shed of Machynlleth in January 1963. On Saturday, 14 November 1964 she is seen at Barmouth with the stock to form the 0945 to Dolgellau. She was more fortunate than the line she was to traverse that day, which was closed the following January, in that she has survived into preservation at the Great Central Railway.

The Cambrian Crusade

Alighting off the Chester-bound train at Ruabon, we caught a southbound Birkenhead to Paddington express, with a lost-looking, Carnforth-allocated Stanier 5MT 4-6-0 45072 at its head, for the short hop to the Italianate-designed station of Gobowen. There, the hoped-for GWR pannier-powered connecting shuttle to Oswestry was disappointingly worked by Ivatt Mogul 46516. Well, at least it was steam, so beggars can't be choosers! This 2½-mile, eight-minute journey was, the following January, turned over to DMUs; the eventual withdrawal of the service in November 1966 awarding Gobowen, by default, the railhead for north Shropshire.

The railway between these two locations might be resurrected one day if the Cambrian Heritage Railway group achieves their aspirations. Similar to the station at Gobowen, Oswestry, albeit now in use as commercial premises, is a Grade II listed building. Once boasting a through road and four bay platforms, this station was the heart of the Cambrian Railways. By the time of our visit, with the cessation of all services, excepting the Gobowen shuttle, being just weeks away there was an air of abandonment and neglect about the place. As for the steam allocation at the shed, with the Ivatt Moguls having saturated the scene for many years, the only remaining GWR representatives were two panniers and two 2884 2-8-0s.

So, as we were about to travel behind the third Ivatt-designed Mogul, a brief word about the class is due. A total of 128 of these 2MT 2-6-0s locomotives were constructed between 1948 and 1953, essentially for light-weight cross-country and local passenger trains. While the majority were built at Crewe, the final twenty-eight, most of which were allocated new to Oswestry, were constructed at Swindon.

Opened in 1849, initially as a branch line terminal from Gobowen, in 1860 Oswestry was reached from the south by the railway from Welshpool, the crossing of the River Vyrnwy being the costliest obstacle. The only habitation of any size along the line was at Llanymynech, a settlement that straddles the English–Welsh border, at which the objective of that afternoon, the Llanfyllin branch, diverged.

The weather that day had by now descended into a cold, rain-sodden scenario and, rather than loiter at Oswestry station for two hours, we travelled on an earlier train through the Llanfyllin junction station of Llanymynech to Welshpool with Ivatt 2-6-0 46509, relying on a three-minute cross-platform 'connection' with an opposite-way working. We knew that, being a single line, our returning train, the 1235 Aberystwyth to

The Chester foreman borrowed the visiting Carnforth 5MT 45072 to work the 42-mile leg of the 1140 Birkenhead Woodside to Paddington service to Shrewsbury. It is seen here arriving into Ruabon.

One of the two remaining panniers allocated at Oswestry, 9-year-old 2F 1668, with just weeks to live, performs carriage-shunting duties. Her stablemate, 1638, was fortunate enough to survive into preservation.

Whitchurch service, couldn't depart without the token that was conveyed by the driver of 46509. Approaching Welshpool, we gathered our bags and cameras and, alighting prior to the train coming to a stand, raced over the footbridge and jumped aboard. Having probably never witnessed a connection such as this being made before, the station supervisor then held up the Whitchurch train, only letting it go after checking our tickets. As we had purchased privilege returns at Oswestry, they were naturally valid; being railwaymen ourselves, any fraudulent travel irregularity would put our employment at risk.

Gathering our thoughts and our breath, after the short nineteen-minute journey back to Llanymynech we alighted to find that one of Shrewsbury's stud of Manors, 7812 *Erlestoke Manor*, had been at the front.

The Llanfyllin branch, opened predominantly to serve several limestone quarries in 1863, saw a further increase in traffic during the 1880s in connection with the construction of the Vyrnwy Reservoir for Liverpool Corporation. The branch trains were, for the first thirty-one

Shrewsbury's 7812 *Erlestoke Manor* departs Llanymynech with the 1235 Aberystwyth to Whitchurch. Erlestoke Manor was in the village of Erlestoke on Salisbury Plain (Wiltshire) and nowadays is in use as a youth custody centre. This station, together with the route from Welshpool to Whitchurch, was to close eight weeks later, with 7812 surviving into preservation at the Severn Valley Railway.

years of operation, accessed by trains having to reverse northwards out of Llanymynech station into the 'Rock Siding' in order to gain height to cross *over* the Ellesmere Canal. Cambrian Railways rectified this anomaly in 1893 by reopening a closed section of the former Nantmawr branch, enabling Llanfyllin trains to start southwards from Llanymynech and cross *under* the Ellesmere Canal.

Shrewsbury's Ivatt 46512 was on that day's Llanfyllin branch duties and is seen after arrival at Oswestry with the 1330 from Llanfyllin. Having worked that branch's last services in January 1965, she was dispatched to Crewe South and withdrawn there in November 1966. After spending several years at Barry, she is now preserved at the Strathspey Railway.

Table 101

LLANYMYNECH and LLANFYLLIN

WEEK DAYS ONLY

Miles		am	pm SX	pm SO	pm	pm	pm SO	Miles		am	am	pm	pm SX	pm SO	pm SO
0	Llanymynech .. .. dep	8 16	..	1255	3 34	3 59	6 30 9 28	0	Llanfyllin .. .. .. dep	7 40	9 45	..	1 30	4 5 4 30	8 15
2	Carreghofa Halt	8 19	..	1258	3 37	4 2	6 33 9 31	1½	Bryngwyn	7 45	9 50	..	1 35	4 10 4 35	8 20
3	Llansantffraid	8 26	..	1 4	3 44	4 8	6 39 9 37	3½	Llanfechain	7 49	9 55	..	1 40	4 15 4 40	8 25
5	Llanfechain	8 31	..	1 8	3 48	4 13	6 43 9 42	5½	Llansantffraid	7 54	10 0	..	1 44	4 20 4 45	8 30
6½	Bryngwyn	8 37	..	1 15	3 54	4 19	6 49 9 45	7½	Carreghofa Halt	8 0	10 6	..	1 51	4 26 4 51	8 35
8½	Llanfyllin .... arr	8 42	..	1 20	3 59	4 24	6 54 9 54	8½	Llanymynech ...... arr	8 4	10 9	..	1 54	4 29 4 54	8 40

SX Saturdays excepted   **SO** Saturdays only

The final Llanfyllin line timetable.

After a forty-five-minute wait, the third train of the day, the 1559 departure, with Mogul 46512 in charge, took us the 8 miles to Llanfyllin. No doubt the Montgomeryshire hills scenery would have been worth viewing but, in the gathering gloom and poor lighting of the single-compartment *suburban* carriage, I will never know. With just four SX (Saturdays Excepted) and five SO (Saturdays Only) trains on the line, it was perhaps unsurprising that patronage was poor, our train conveying just a couple of other passengers. Retracing our steps to Oswestry, we then caught our fifth Mogul, 46514, which was working the 1725 departure to Whitchurch. Passenger services over the 'main line' through Oswestry, together with the Llanfyllin branch ceased on 18 January 1965.

We then made our way to Crewe, where our 1854 Euston service arrived with Type-4 D314 emitting loud backfiring-type noises, only to be replaced by sister D342, the latter getting into London a mere sixteen minutes late at 2201. Thus, I completed my longest and most exhausting trip so far: a total of 190 steam miles behind eleven different locomotives from seven classes over 114 miles of condemned lines. This hobby of mine was becoming addictive, visiting parts of the country I would never have considered if it wasn't for the combination of Dr Beeching and the imminent demise of steam.

The Great Western Steam Retreat

Time arr	dep	Station	Traction	Date/Shed
	2235	Euston	D316	Fri 13
0132	0219	Crewe	D227 *Parthia*	Sat 14
0435	0520	Bangor	41234	6H
0655	0735	Pwllheli	82020	6F
0912	0931	Barmouth	80104	6F
0936	0954	Morfa Mawddach	46521	6F
1015	1049	Dolgellau	75023	6C
1241	1258	Ruabon	45072	10A
1312	1320	Gobowen	46516	6E
1328	1418	Oswestry	46509	6E
1455	1458	Welshpool	7812 *Erlestoke Manor*	6D
1517	1559	Llanymynech	46512	6E
1624	1630	Llanfyllin	46512	6E
1711	1725	Oswestry	46514	6E
1812	1826	Whitchurch	DMU	
1844	1854	Crewe	D342	
2145		Euston		

6C – Croes Newydd; 6D – Shrewsbury; 6E – Oswestry; 6F – Machynlleth; 6H – Bangor; 10A – Carnforth.

10

Bashing to Banbury

So now we enter 1965 – the final full year of steam in the Western Region. On 1 January there were 4,990 steam locos in BR stock, a decrease of 2,084 over the previous year, and by the year's end the number would be further reduced by 1,987.

Out of the remaining steam motive power depots within the WR, the first casualty was Reading, which closed on 4 January with, surprisingly, none of the seven occupants being condemned – her three 61xxs being dispatched to Southall and two Halls each to Didcot and Worcester. This left just twenty-two sheds, with an allocation of 521 locomotives between them, remaining on the Western Region's books. Of these, just over 400 were of GWR origin, the balance being made up of former LMS Ivatt Mickeys, Fowler 4Fs and Stanier 8F 2-8-0s, together with four variations of BR Standard classes. However, a further 154 ex GWR locomotives were allocated to sheds in Western Region's former Wolverhampton Division (84 group), those having been transferred en bloc to the LMR with the 1963 boundary changes. Furthermore, routine steam locomotive repairs at Swindon ceased in February, with Ivatt 2-6-0 'Flying Pig' 43003 being the last out shopped.

Three sheds were to lose their steam allocations in March: Aberdare, Treherbert and Old Oak Common. The two Welsh sheds had their handful of remaining tank locomotives dispersed among their neighbours and, while eight of the twenty-two Pannier tanks at the once mighty and glorious Old

Oak Common were condemned, eleven escaped to Southall, with one each going to Oxford, Bristol Barrow Road and Bath Green Park.

Other railway news from the early months of that year saw Sir Winston Churchill's state funeral train travel from Waterloo to Oxfordshire and the second Beeching report, *The Development of Major Trunk Routes*, proposing which lines should receive investment, was published. This envisaged a reduction of 50 per cent of the existing 7,500 route miles and outlined those deemed worthy of investment.

Personally, upon realising the significance of the rapidly disappearing steam locomotive, starting from mid-February and coinciding with increased income from promotion within the BR clerical grades, I was to spend almost every weekend of the following three years careering around the UK after steam haulages. The weather was irrelevant. Come snow, heatwave or rain, nothing stopped the frantic pursuit of steam before it was too late.

It is difficult to remember the reasons for certain actions when attempting to recall events more than half a century later, but on three separate occasions during spring 1965, I enacted exactly the same 270-plus steam miles train plan, visiting Banbury then Salisbury, on the same five trains. Retrospectively, I believe the attraction was, in addition to attempting to travel behind all the remaining Bulleid Pacifics, the 22¾ miles of former GWR motive power between Banbury and Oxford on the southbound York.

Being 18 years of age that year, I failed to appreciate the speed at which the WR had dieselised all other main lines and that I had missed many opportunities to travel behind former GWR motive power. With my steam travels in their infancy, I began, knee-jerk fashion, to home in on areas where it was disappearing earlier rather than later. This pocket of GWR steam at Oxford and Banbury enabled me to obtain runs behind ten examples of Hall and Grange class 4-6-0s, being much more appreciated by me than if not at all. It must have been a sad scene to witness for older and more travelled enthusiasts, but to me it was a last lifeline eagerly grabbed before steam's untimely but inevitable demise.

The northbound Pines Express was diagrammed for a Bournemouth-allocated Bulleid Light Pacific to Oxford, a Brush Type 4 (Class 47) to Crewe – with electric traction on the two portions to Manchester and Liverpool. The turnaround time at Oxford (1248–1404) was sufficient for the 70F-allocated locomotive to be serviced and turned for the southbound equivalent. The Bournemouth to York, however, was not so cost-effectively resourced. While the Eastleigh-based locomotive was able to replicate the Pines Express duty in that she could also work to and from Oxford (1403–1521) on a balanced duty, and the ER-based Type 3 (Class 37) was able to work to and from Banbury (thirty-five-minute turnaround) via the ex GCR line, the power over the link between Oxford and Banbury was left to the WR to resource. Not only was it just a 22¾-mile stretch of line, but also – to compound the rostering requirement – the trains actually crossed en route. This scenario, apart from the economics of it all, frustrated the enthusiast in that he could only obtain runs with just one of the two WR-allocated locomotives in use that day. All the above was related to Saturday diagrams; I was never privy to the SX workings.

Catching the Pines Express from Basingstoke, and having passed through Oxford and noted the cramped shed packed with GWR power, we then travelled over the line that these days is marketed as the Cherwell Valley Line. Unlike many modern-day brandings, this actually is an appropriate description as it intertwines with the River Cherwell itself, and indeed the adjacent Oxford Canal, for the majority of its length. Completed in 1850,

During the hour's wait between trains at Banbury on Saturday, 27 March 1965, this and the following two scenes were taken. At 26 years old, 6874 *Haughton Grange* (named after a late nineteenth-century Cambridgeshire house allegedly used as a government animal-testing centre in the 1960s) storms through with a southbound train of empty car flats.

An unusually clean 6917 *Oldlands Hall* works through her home station with a northbound freight while, in the up bay, Worcester's 6848 *Toddington Grange* waits to work the 1008 York to Bournemouth West, the 22¾ miles to Oxford.

BR 9F 92133, a recent transfer from Leicester to Birkenhead the previous month, at 7 years old, heads through with a northbound train of cars from Cowley.

the countryside through which the valley passes makes a pleasant break between the university city of Oxford and the sprawling expanse of Banbury.

With, until 1966, locomotive depots located at Oxford and Banbury, together with the route being utilised for cross-country services, both these establishments were natural locomotive- and crew-changing locations. I had anticipated the one-hour dwell time at Banbury was going to be boring, but there was never more than a ten-minute gap between steam movements with several freight trains running on Saturday afternoons.

Banbury station had been opened by GWR in 1850, with a gradual expansion of platforms and passing loops being added as traffic levels built up. This increased even further upon the connection at Banbury North with line from the GCR at Culworth Junction opening in 1900. Its status was further enhanced ten years later when the cut-off via High Wycombe was opened. It was rebuilt extensively in 1958 into the modern architectural establishment I witnessed the year of my visit.

Time arr	dep	Station	27/3/65	03/4/65	24/4/65
	0954	Waterloo	34026	34026	34005
1114	1142	Basingstoke	35011	35023	34047
		Oxford	D1711	D1685	D1701
1325	1433	Banbury	6848	6903	6990
		Oxford	34097	34036	34009
1630	1648	Basingstoke	34057	34057	34048
1800	1835	Salisbury	34089	34089	34052
2021		Waterloo			

At the top of the charts on the first of my outings was 'The Last Time' by the Rolling Stones, ousted just weeks later by the Beatles' 'Ticket to Ride'. Both titles were loosely appropriate to the many steam journeys undertaken that year. As for footie fans, Manchester United won that year's First Division title and Liverpool won the FA Cup, beating Leeds United 2-1.

Appetite whetted, I returned on Friday, 7 May for the northbound York–West Country 34097 *Holsworthy*, giving way to 6947 *Helmingham Hall* at Oxford then at Banbury to 'Tractor' D6817. Rather than alighting at

Banbury for a two-hour fester for a returning southbound DMU, I elected to travel through to Nottingham Victoria, essentially in order to travel the 126½ miles on the 1715 to Marylebone (required steam tracks Nottingham to Loughborough).

My next visit to Banbury was on the penultimate Friday of the last remaining booked steam-operated service from London Paddington, the 1615 departure. I was aware that on the final day, 11 June, the train was to be worked by 7029 *Clun Castle*, but I was already committed to a family holiday in Italy. Twenty-two-year-old class 4-6-0 6952 *Kimberley Hall* was the motive power for my journey and a leisurely two-and-half-hour, 67½-mile ride was enjoyed. We were looped at Princes Risborough for the 1650 Birmingham Pullman to pass and at Bicester North for the 1710 Paddington to Wolverhampton. After the commuters had alighted at High Wycombe there were few other passengers or even enthusiasts aboard, a scene no doubt *not* replicated a week later.

Springtime is usually regarded as beholding a great summer but for those steam locomotives allocated to the WR the cutter's torch was harvesting vast annihilation throughout the land. While the number of ex GWR locos

Shrewsbury shed hosts Oxford's 6947 *Helmingham Hall* (a moated Suffolk manor once used as a location for the BBC's *Antiques Roadshow*) in light steam, five months after my trip with her.

This train, the 1615 Paddington to Banbury, had become the final booked steam departure out of Paddington. One week prior to its cessation, Friday, 4 June 1965, and Banbury's 6952 *Kimberley Hall* is seen being signalled into the down loop at Bicester North in order for the 1710 Paddington to Wolverhampton to pass.

A lost-looking Burton-allocated LMS 8F 2-8-0 48117 draws out of Banbury yard with a southbound freight.

within the former 84-coded group (now LMR) weren't faring too badly, the WR steam allocation had dropped to just over 400 in three months.

April saw the closure of Rhymney (a sub-shed of Merthyr), with the majority of its 56xx 0-6-2Ts being condemned, and Pontypool Road, its allocation being dispersed locally. Then it was Didcot's turn – the *Oxford Mail* running an article depicting the final steam departure on Sunday, 13 June comprising 7816 *Frilsham Manor* (en route to her new home of Gloucester) hauling two dead Halls to Oxford.

June saw the demise of the once vast Exmouth Junction shed. Three of its four Panniers were condemned and 9647, along with six Ivatt 'Mickeys' and two Standard 4MT tanks, being dispatched to Templecombe. Yeovil, itself closing just weeks later, received one 75xxx and a Standard 4MT tank, while Worcester benefitted from three 75xxxs, and four 82xxx 3MT tanks went to Gloucester.

Neath also closed that month, and of her fifteen tanks, seven were condemned. The remainder, presumably the fittest, were dispatched to Cardiff East Dock, Radyr or Llanelly for a short extension of their lives.

11

The Inter-Regional Finale

We now move on to the summer of 1965, when the top spot of the charts was monopolised for five weeks by either 'I'm Alive' by the Hollies or Elvis Presley's 'Crying in the Chapel'. There were now fourteen sheds retaining an allocation of just over 220 steam locomotives throughout the Western Region. Cardiff Radyr was the next to go, and of her eleven tanks only two escaped: one each to Severn Tunnel Junction and Newport Ebbw Junction.

While the frequency of the inter-regional passenger services in the 1960s does not compare with the clockface pattern of today's modern railway, the supplemented Saturday timetable during the summer period managed to transport thousands of holidaymakers from all points throughout the north and Midlands to and from the south coast resorts. This was, of course, in the days of only one (if you were fortunate) car per family and when there was very little of the present extensive motorway network.

The main objective of an overnight trip to Weymouth on Saturday, 17 July (days after the Great Train Robber Ronnie Biggs had escaped from Wandsworth) was to travel on the 1105 Weymouth to Wolverhampton train. This short-dated Saturdays-only train was routed over the by then steam-starved routes via Yeovil Pen Mill, Swindon and the rarely used (by passenger services) Didcot West Curve.

At 0515 on Saturday, 17 July 1965, I visited Templecombe shed. Wandering around (unchallenged), an air of neglect and abandonment seemed to prevail – the silent locomotives dripping with early morning mist. This photograph depicts a line of withdrawn locomotives already condemned and awaiting removal to the breakers: 9647, 9670, 41214, 80067, 41243 and the still in service 41296 at the rear.

Starting out from London's Waterloo on the Down TPO (2235 departure), I could easily have gone through to Weymouth and festered there for several hours, but as I was always on the lookout to encompass as many runs behind different steam locomotives as possible, I alighted at Eastleigh and travelled across via Salisbury to Templecombe, arriving there at 0500.

There was plenty of time while awaiting the start of services that morning and so, without being challenged, I bunked the shed.

17/7/65	83G Templecombe @ 0515
In steam	41283, 41290, 41291, 41307, 48706 (82F), 75072, 80037, 80041, 80043, 80059 (82F)
Dead	41216, 41296
Withdrawn	9647, 9670, 41208, 41214, 41243, 80067

Resuming my journey to Weymouth, I headed south over the S&D on the 0735 departure hauled by 83G's 4MT 4-6-0 double-chimney 75072 to Poole, en route crossing its other stablemate, sister 75073, on the 0710 out of Bournemouth West. From Poole former WR green-liveried BR Standard 5MT 73029 took me into Weymouth, where its sister 73020 sat waiting with the objective of the day.

The Wilts, Somerset and Weymouth Railway obtained its assent for construction in 1845 but, following a series of revised route proposals, did not open for business until 1857, with mixed gauge south of Dorchester to accommodate the L&SWR services from Southampton. Converted to standard gauge in 1874, the line was a difficult one to work with the gradients of both Evershot and Bincombe banks to master. The Channel Isles Boat Train ran this way from Paddington until 1959, after which date the SR took the reins.

Oodles of new steam track, indeed new on any form of traction, now lay ahead of me. The 13-year-old Weymouth-allocated Standard 5MT was fortunately up to the task, which was just as well as ahead lay the 4-mile climb (as steep as 1 in 50 at one point) up to Bincombe Tunnel, followed a few miles later by the near 4-mile slog up to the summit of the line at Evershot. Indeed, all around knew she was coming from miles away!

Then we hurtled down the 1 in 51 to Yeovil before joining the GWML from Castle Cary to Westbury before swinging north via Melksham to join the Bristol line at Thingley Junction. I was thoroughly enjoying this ride through the diesel desert that Wiltshire had become.

On we went through Chippenham, joining the South Wales line at Wootton Bassett. We called at Swindon (noting the only steam sighting en route of the withdrawn ex Bromsgrove banker 0-6-0PT 8405) before then taking the Didcot west curve and heading north to Oxford, thus completing a wonderful three-and-a-half-hour 125-mile journey.

After arriving into Oxford, rather than seeing what was about, i.e. GWR-powered services, I dashed through the subway and, after purchasing a ticket back to SR territory and hearing announcements to the effect that the train standing in the up platform was for Basingstoke, I jumped aboard. After asking passengers where the train had originated, I worked out that it was the thirty-three minutes late-running 1025 Manchester Piccadilly to Bournemouth West, which should have departed at 1431 – effectively making a minus twenty-four-minute connection. Only upon alighting

Table 13									Saturdays

20 June to 5 September

Wolverhampton and Birmingham to Yeovil and Weymouth

WOLVERHAMPTON L.L.	d	6 25			11b10			16 15	
BIRMINGHAM SNOW HILL	d	7b05			11b34			16 28	
LEAMINGTON SPA GENERAL	d	7 41			12 03			16 51	
BANBURY	d	7 27			12 34			17 22	
OXFORD	d	8 37			13 10			18 18	
CHIPPENHAM	d	9c30			14c01				
TROWBRIDGE	a				14 23			19 02	
WESTBURY	a				15 03			19c27	
YEOVIL PEN MILL	a	11 00			15 14			20 18	
DORCHESTER WEST	a	12 14			16 23			20 53	
WEYMOUTH	a	11 47			17 05			21 20	
					17 17			21 57	
								22 58	

WEYMOUTH QUAY	d					16 15		
WEYMOUTH	d	11 05				16 28		
DORCHESTER WEST	d	11 25				16 51		
YEOVIL PEN MILL	d	12 02				17 22		
WESTBURY	d	12 54				18 18		
TROWBRIDGE	d	13 05						
CHIPPENHAM	d	13 39				19 02		
SWINDON	d	14f17				19c27		
OXFORD	a	15 00				20 18		
BANBURY	a	15 37				20 53		
LEAMINGTON SPA GENERAL	a	16 04				21 20		
BIRMINGHAM SNOW HILL	a	16 39				21 57		
WOLVERHAMPTON L.L.	a	17 12				22 58		

Heavy figures indicate through carriages
For general notes see page 49

b Passengers travelling by this train beyond Banbury are required to hold Regulation tickets see page 45
c Arr 3 minutes earlier
e Arr 5 minutes earlier
f Arr 14.01

Oxford was always a busy location on a summer Saturday, with a plethora of inter-regional trains changing locomotives. Here Cardiff East Dock's 6815 *Frilford Grange* makes her way to the shed (the wooden shack on the left) having just brought in the 1025 Manchester Piccadilly to Bournemouth West, the train from which I took the photograph.

at Basingstoke was I able to note what was at the front: Merchant 35011 *General Steam Navigation* – no former GWR power that day!

That July, and indeed for a nine-week period that summer, the charts were ruled by either 'Mr Tambourine Man' by the Byrds, 'Help!' by the Beatles or 'I Got You Babe' by Sonny and Cher. Other non-railway news was that cigarette advertising was banned on TV and the Wimbledon tennis champions that year were both Aussies, Roy Emerson and Margaret Court.

Working a Saturday morning was part and parcel of a BR clerk's week back then and, having fulfilled that obligation, I headed for Basingstoke (1242 departure out of Waterloo with 34050 *Royal Observer Corps*) on Saturday 31 July to board the 1311 Portsmouth Harbour to Wolverhampton Low Level. Brought in on time by Standard 5MT 73016, I suffered an eventual eleven-minute late departure, for reasons not documented at the time, with Oxley's Stanier 5MT 45263. Although disappointed that it wasn't a GWR workhorse, I decided to stay aboard as I required the track north of Banbury for coverage with steam. With so many 'extra' trains along the route, time-keeping on all my visits that summer was disappointing to say the least. With a signal failure at Aynho Junction and the need for an extra water stop at Leamington Spa, I experienced an eventual seventy-eight-minute late arrival into Wolverhampton, at 1850. Noting the line was awash with steam and after vowing to return ASAP after a forty-minute wait, I caught the final Birkenhead-emanating train of the day into Paddington and arrived at 2215.

The only shed closure that August was Cardiff East Dock. From an enthu-siast's viewpoint, this shed gained importance upon the closure to steam of the nearby Cardiff Canton in September 1962 by inheriting the majority of its steam locos. A significant number of these were subsequently dispatched to Dai Woodham's yard at Barry and ended up in the hands of preserva-tionists. By the time of East Dock's shed closure, however, just twenty-two locomotives remained, nine of which were immediately condemned. Of the others, Severn Tunnel received four, Gloucester two, Oxford one and Newport Ebbw Junction six, including the two surviving BR 9Fs.

Choices had to be made by steam followers that summer of 1965 and, with Scotland being prioritised in the hopes of runs with both the A2 and V2 classes, it wasn't until late August that I once again returned to the Oxford line. Short-dated summer Saturday-only services (those depicted with wavy lines in the timetable) from south coast resorts to the Midlands and north-east were plentiful over this former GWR route from Reading to Birmingham and, with the majority of them being resourced with steam, they attracted many an enthusiast hoping for runs with GWR locomotives.

So, on the final Saturday that August, having once again worked my Saturday morning duty in my London office, I headed for the inter-region-als, in this instance by travelling out of Paddington, noting Barrow Road's

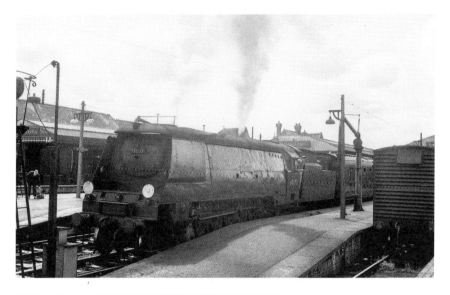

Basingstoke on Saturday, 31 July and Unmodified WC 34033 *Chard* awaits the off with the 1229 (Relief) Portsmouth & Southsea to Wolverhampton Low Level. As this train was booked for DL haulage north of Oxford, I let her go, catching the main Portsmouth train, which was worked from Basingstoke with a Black 5MT. I never did get a run with *Chard*!

In full flight with a southbound freight at Priestfield (Wolverhampton), Bristol Barrow Road's Modified Hall 7907 *Hart Hall* is captured in camera from my passing train.

7924 *Thornycroft Hall* just outside the station, with the intention of inter-
cepting them at Reading West. This day was to be yet another example of
chaotic timekeeping. The train I was aiming for was the 1358 departure out
of Reading West. After witnessing Oxford's 6923 *Croxteth Hall* crawling tan-
talisingly through with a late-running northbound service, what did my train
turn up with? Brush Type-4 D1683, that's what! With no back-up plan, I had
no option other than to board it and, having suffered extended delays south
of Oxford (due to preceding trains changing locomotives), I eventually had
success, in my bid for runs with GWR locomotives, with a decrepit-looking
6910 *Gossington Hall* on a 1614 Wolverhampton departure out of Oxford.

Indeed, the majority of GWR locomotives were in a similar condition,
with their brass cabside numberplates and names already removed, together
with their smokebox numberplates; ostensibly, it was alleged, this was to
thwart souvenir hunters among the enthusiast community. It has also been
claimed that the removal of such adornments was a manoeuvre by management, who classified them as unnecessary beautifications that resulted in the
locomotive becoming emasculated, although some depots made an effort to
stencil the number on the cabside in a somewhat amateurish fashion. That

It's the final Saturday that August, the 28th, and the usual signal delays outside Oxford
allowed me to take this shot of WC 34001 *Exeter* departing with the 0835 Leeds City to
Poole, a train she had taken over at Oxford.

Every movement was noted and photographed by hundreds of enthusiasts during that last summer of GWR steam. Bristol Barrow Road-allocated, 15-year-old Modified Hall 7914 *Lleweni Hall* departs Oxford with a southbound afternoon service.

While waiting at Poole on Saturday, 9 May 1964 for a Weymouth-bound train, Oxford-allocated 4-6-0 6910 *Gossington Hall* (a large house in Gloucestershire that was the setting of an Agatha Christie novel) passed through on what I wrote in my notebook at the time as 'a lightweight parcels train'. I was fortunate enough to catch a run with her into Birmingham just five weeks before her demise.

may well have been the case but, with just a chalked scrawled number on the cabside as a locomotive's sole means of identification, observational mistakes were likely to be made.

On this occasion, as if to vindicate the Hollies' summer 1965 hit, 'I'm Alive', *Gossington Hall*'s internal condition must have been reasonable because, after barking up Hatton bank with her exhaust clouding the sultry Warwickshire countryside, she lost a mere six minutes on the near two-hour run to Birmingham. Upon arriving into Snow Hill, and having espied a Paddington-bound train awaiting departure, I ran hell for leather over the footbridge to board it, reaching home two hours earlier than my previous foray along the line.

I had 'done' Scotland that summer and the Southern could wait, so for the first two Saturdays in September I again blitzed the Oxford services. If time permitted and I was not working, I always chose the steam services from Waterloo to Basingstoke to connect into the cross-country services rather than taking a DL-worked train out of Paddington. At Basingstoke the 1042 ex Poole, which had a Hall on it the previous week, on this occasion had

This, together with the following seven photographs, were all taken on Saturday, 4 September – the last day of the 1965 summer service. An unusually clean BR 9F, 92094 of Birkenhead, waits the signals at Oxford. She escaped the mass cull upon 8H's closure in November by moving to Speke Junction, until that shed's closure in May 1968.

BoB 34060 *25 Squadron*, and although at least it was steam, I would have appreciated GWR power! Alighting at Oxford in the hope of a GWR-powered northbound train, the time was passed pleasantly with a steamy scenario soon to vanish forever, with four differing classes being observed. Firstly, 6803 *Bucklebury Grange* departed with a southbound train, then WD 2-8-0 90295 came through with a northbound freight. No. 6967 *Willesley Hall* then shuffled around on station pilot duties and finally Standard 9F 92094 posed for photographs in the centre road.

The wait was worth it. Oxford-allocated 6932 *Burwarton Hall*, having taken over at Basingstoke, was working that day's 1110 Bournemouth West to Newcastle Central. I travelled with her to Banbury, alighting there and expecting a mere fifteen-minute wait for the 1105 Weymouth to Wolverhampton. This eventually arrived twenty-eight minutes late with a previously travelled with Black 5MT 45263. Alighting off the now forty-six-minute late-running 1640 arrival Birmingham Snow Hill – and contemplating returning to London – I was dissuaded when 6849 *Walton Grange* arrived with the 1328 ex Portsmouth Harbour, which I took to its Wolverhampton destination and returned into London once again on the 2210 arrival.

Twenty-three-year-old 6932 *Burwarton Hall* calls for water replenishment at Oxford. She was working that day's 1110 Bournemouth West to Newcastle between Basingstoke and Banbury.

A further batch of seventy-one Halls were constructed between 1944 and 1950 to a Hawksworth redesign specification, becoming known as Class 6959 Modified Halls. Here, at Banbury, 6980 *Llanrumney Hall* acts as the station pilot. Llanrumney Hall is a listed 1450-built structure in the Cardiff suburbs that over the years has been a remand centre and a pub, and is allegedly haunted by the remains of Welsh royalist Llewellyn the Last, who was beheaded there!

Disappointingly (I was hoping for GWR power), Oxley's 45263 arrives into Banbury, with the 1105 Weymouth to Wolverhampton Low Level having taken over at Oxford.

You had to be quick to record the individual number of the ex GWR locomotives back then, with usually just a chalked or stencilled indication of the identity on the cabside. In a typically rundown condition, Oxley's 6833 *Calcot Grange*, withdrawn the following month, heads south through Birmingham Snow Hill.

7812 *Erlestoke Manor* arrives light engine at Wolverhampton Low Level. At 26 years old, she was withdrawn two months later at her home shed of Shrewsbury in the Manor cull (see Chapter 13).

Seen during her short four-month tenure at Oxley is one of the three Brits drafted in to work the West of England services via Bristol that summer. BR 7P6F 70053 Moray Firth arrives into Wolverhampton Low Level with the 1110 from Ilfracombe. Weeks later, she was dispatched to Banbury for use on the ex GCR services into Marylebone, before ending her days at Kingmoor in April 1967.

Derby-built, 12-year-old, Riddles-designed 4-6-0 Standard 5MT 73036 is seen having arrived into Wolverhampton Low Level on an unidentified train, probably originating from the West Country.

Wolverhampton Low Level opened in 1854, initially as Wolverhampton Joint before being renamed Low Level two years later; the adjacent LNWR station of Queen Street being retitled High Level at the same time. Converted to standard gauge in 1869, there was once an overall roof but this suffered corrosion and it was taken down in 1922, with GWR installing standard platform canopies. With the cessation of the Paddington to Birkenhead express in 1967 and the diversion of DMUs from Shrewsbury and Stourbridge into Birmingham New Street, the paltry remaining services (a direct comparison can be made with Nottingham's Arkwright Street station) were withdrawn in 1972. It was used for many years as a parcel concentration depot but the Grade II listed main building has found a new lease of life as a banqueting hall and wedding venue.

Back again the following week, the iconic '(I Can't Get No) Satisfaction' by the Rolling Stones hitting the airwaves, I was able to obtain a return steam trip to Banbury with GWR power. Casually looking through that week's SR (SWD) Special Traffic Notice, I observed that a 0958 Relief train Bournemouth Central to Wolverhampton was running that Saturday. With the arrival time into Banbury of 1309, an easily made connection into the southbound York of 1433 beckoned: and you never know, it might be steam-hauled north from Oxford!

After travelling down on the 0930 out of Waterloo with 73113 *Lyonnese*, the relief turned up with BoB 34077 *603 Squadron* and this gave way at Oxford to Modified Hall 7909 *Heveningham Hall*. I was not normally a shed basher but as the weather had now changed from rain to dry and sunny, on arriving at Banbury I thought 'why not'. Following the directions as detailed in Fuller's ever-present *Directory* (primarily purchased for its street maps of northern cities), I walked the walk to 2D (formerly 84C):

The Shed was on the west side of the line south of Banbury station. Turn left out of Banbury station into a narrow road running parallel to the railway. This turns into a cinder track and leads to the shed. Walking time ten minutes.

The Inter-Regional Finale

11/9/65	2D Banbury @ 1400
In steam	6697, 6934 *Beachamwell Hall*, 6951 *Impney Hall*, 6964 *Thornbridge Hall*, 6976 *Graythwaite Hall*, 44847 (16B), 45052 (2J), 48010 (1A), 48056 (8A), 48220 (2E), 75004 (6F), 92047 (8H), 92059 (8H), 92067, 92074, 92224
Dead	6671, 6930 *Aldersey Hall*, 34051 *Winston Churchill* (70E), 73048, 92128, 92213, 92228
Withdrawn	6644 (2A), 6911 *Holker Hall*, 6916 *Misterton Hall*, 6922 *Burton Hall* (2A), 44560 (85B)
Diesels	D1715, D3709

This and the following seven photos were all taken on Saturday, 11 September 1965 when bunking Banbury shed. This one depicts a very healthy steam scene featuring 75004, 48010, 45052, 6934 *Beachamwell Hall* and 48220 in the line-up. While the WR rid itself of steam by the year end, Banbury remained open until October 1966 due to the necessity of dealing with visitors off the SR together with the ex GCR requirements.

Although displaying a chalked or painted 2D on her smokebox, this 37-year-old 0-6-2T 6644 was, on paper, withdrawn at Tyseley two months previously.

This Banbury shed scene sees resident 0-6-2T 6697, a locomotive that was to be transferred to Croes Newydd by the end of the year. The move ensured its longevity as one of the last two members of its class. There is a surprising visitor in the form of BoB 34051 *Winston Churchill*. Both locomotives survived into preservation, albeit statically.

Gloucester-allocated, Fowler-designed, Armstrong Whitworth-built, 43-year-old 4F 44560 reposes in the sun, this former S&D locomotive having been withdrawn just days earlier.

Willesden-allocated 2-8-0 Stanier 8F 48010 at rest. She was to end her days at Newton Heath in January 1968.

This 1942 North British-built 2-8-0 48220 is raring to go, home to her Saltley depot perhaps?

Machynlleth's BR 4MT 75004 visits on the day. I was to view her again in her own territory working a freight the following month and finally caught a run with her in September 1966 along the Welsh Marches line.

Once working the famed GCR *Windcutters* when allocated at Annesley, this now Banbury resident BR 9F 92074 is under the coal chute.

The southbound York had 6951 *Impney Hall* on it, relinquishing the train at Oxford to Standard 5MT 73022. Going through to Southampton with her and catching a required Standard 4MT 80146 on a local, that day's steam mileage amounted to a healthy 287½ miles behind eight locos (six required). The eleven hours of steam cost (at privilege rate, mind) a hefty 18*s* 6*d* (82½p), an excellent albeit expensive day's outing.

The Paddington management were hell-bent on achieving the accolade of becoming the first region to implement the BRB's diktat regarding the elimination of steam power. They had announced, during autumn 1965, that from midnight on 3 January 1966 no facilities for servicing steam locomotives would be available and that their own allocation would be completely withdrawn. While this was subsequently thwarted by the deferment of the Somerset & Dorset route that resulted from the bus companies charged with providing the replacement bus arrangements pulling out at the last moment, the end came a mere eight weeks later and the WR management had achieved its goal.

12

A Saturday Somerset Sojourn

With the closure date for the S&D having been announced, I made what I thought was to be my final visit to the line that September, having realised that I had yet to travel over the gradient-strewn Mendips section between Bath and Evercreech Junction. Regrettably, I was too late to witness the line during its summer Saturday heyday when double-headed inter-regional trains from a wide variety of northern and mid-England destinations traversed it to access the south coast resorts of Poole and Bournemouth. In September 1962, with the diversion of the famed Pines Express, it was relegated to a secondary route and left to wither and die.

By September 1965, the steamiest overnight route by which to access the S&D was using the 2235 mails from Waterloo to Eastleigh and changing there onto a 0202 Hymek DL-powered departure for Bristol. Having arrived there just before 0500, to fill the hours prior to the Bath Green Park departure of 0750 I track-bashed the 13½-mile-long Severn Beach branch. After the uninspiring scenery I passed through, coupled with nearly two hours of fume-ridden DMU occupancy, I couldn't have been more pleased to see BR 3MT 82041 sitting on my selected train. In hindsight, I was indeed fortunate with this catch as both the 0600 and 0900 trains over the same route were Hymek DL operated. Glimpsed en route to Bath was Bristol's Barrow Road (82E) depot, which had two Black 5s, one LMS 8F,

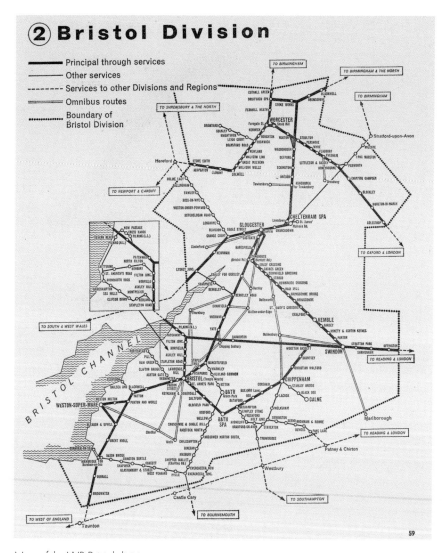

Map of the WR Bristol division.

one Pannier tank and the subsequently preserved 4920 *Dumbleton Hall* in situ. Standard 3MTs 82030 and 82044 were also noted in steam in the area.

Having nearly an hour to wait for my southbound train, I bunked the shed at Bath Green Park. Although I took a shot of the Jinty hiding in the depths of the shed, its poor quality precludes its use here.

It's Saturday, 18 September 1965 and 25-year-old 0-6-0PT 3659 rests outside Bath Green Park shed. On paper she was allocated to Templecombe but was withdrawn at 82F the following month.

Busying herself on the same day, resident sister 3758, whose career had included lengthy stays at both Bristol sheds of St Phillips Marsh and Barrow Road, is seen near her home depot of Bath Green Park.

Eleven-year-old Standard 5MT 73068 being prepared for the arduous climb over the Mendips at her home shed of Bath Green Park. She was to work the 0900 ex Bristol Temple Meads the 60 miles forward to Branksome with me aboard.

18/9/65	82F Bath Green Park @ 0900
In steam	3659 (83G), 3758, 48309, 48444, 73001, 73068, 80059, 82041
Dead	47506, 48706, 48760, 82004
Withdrawn	48737, 73015, 73051, 73054

What a superbly engineered line this was. First, there was the awesome 1 in 50 climb affording spectacular views over the City of Bath before plunging headlong into Devonshire tunnel, and then through to the summit in what was Britain's longest unventilated tunnel (1,829yd) at Combe Down. I believe crews had to cover their faces with wet clothes to avoid asphyxiation when travelling through this tunnel. Nowadays, it is incorporated into an LED-lit shared walking/cycle pathway. Then, after a short run down into Midford, 8 more miles of 1 in 50/60 was encountered to the summit of the Mendip Hills at Masbury (811ft).

A Saturday Somerset Sojourn

As if that wasn't enough, there were then 7 miles of downhill switchback to Evercreech through splendidly unspoilt countryside – all so adequately encapsulated by Ivo Peters and his camera. I feel sorry for those who never had the opportunity to appreciate the enchanting vista I enjoyed that day. 'Window hanging', a favourite activity often undertaken by steam train aficionados such as myself, was necessary in order to drink in the Standard 5MT 73068's efforts. Smothered by cinders, deafened by the exhaust echoing off adjacent rock faces – this journey had it all.

All too soon it came to an end and at Templecombe I returned with (Feltham-allocated?) Standard 4MT 80085 to Evercreech Junction in order to connect once again onto the 1315 Highbridge departure. The new order of Ivatt tanks had replaced the ex GWR power and 41223 scuttled over the Somerset Levels, arriving into Highbridge at 1413. A deliberately uneconomic train plan was then observed with a fresh locomotive and coaching set starting out of Highbridge just five minutes later with sister 41307. As a retired train planner, I despair at this arrangement! If one or the other trains had been retimed by just five minutes the same crew/loco/vehicles could have been

Transferred in from Exmouth Junction that May, 83G-allocated Mickey 41307 departs Highbridge with the 1418 to Templecombe. I had arrived five minutes earlier with sister 41223 and had raced back up the line to take this shot, looking over my shoulder as I ran.

used – but then again BR wanted shot of the line and any cost savings accrued through such a move might have put the closure plan in jeopardy.

I now headed east to Chippenham, where it was the final day of the 5¼-mile Calne branch. Opened in 1863, for many years passenger traffic was never the main income but instead the conveyance of sausages, pies and meat products from the Calne factory of Harris & Company. Although not the last trains, the 1756 out and 1817 return services were well filled with locals

Table 44

Weekdays

NOT on Saturdays 20 June to 5 September

Chippenham and Calne

																								SO
CHIPPENHAM .. .. .. d	6 05	..	8 07	8 53	..	9 55	..	11 55	..	13 40	14 55	15 36	..	16 31	17 12	..	17 56	18 37	..	19 33	21 12	22 55	..
STANLEY BRIDGE HALT d	6 11	..	8 13	..		10 01	..	12 01	..	13 46			..	16 37	17 18	..		18 43	..	19 39	..	23 01	..
BLACK DOG HALT .. d	†		†	†		†		†		†	†	†		†	†					†		†	
CALNE .. a	6 20	..	8 22	9 06	..	10 10	..	12 10	..	13 55	15 08	15 49	..	16 46	17 27	..	18 09	18 52	..	19 48	21 25	23 10	..

Saturdays

20 June to 5 September

																		A		A								
CHIPPENHAM .. .. d	6 05	..	8 07	..	8 55	9 55	..	11 55	12 38	13 42	..	14 55	15 42	16 31	..	17 12	18 00	18 37	19 25	20 42	21 18	22 55	..
STANLEY BRIDGE HALT d	6 11	..	8 13	..		10 01	..	12 01			..		16 37		..	17 18	..	18 43	19 31	..	..	23 01	..
BLACK DOG HALT .. d	†		†					†					†			†		†	†			†	
CALNE .. a	6 20	..	8 22	..	9 08	10 10	..	12 10	12 53	13 55	..	15 08	15 55	16 46	..	17 27	18 13	18 52	19 40	20 55	21 30	23 10	..

Weekdays

NOT on Saturdays 20 June to 5 September

Calne and Chippenham

CALNE .. .. .. d	7 05	..	8 26	..	9 15	11 10	..	13 10	14 05	..	15 11	..	16 02	..	16 50	17 32	..	18 17	19 05	..	19 55	21 35	..
BLACK DOG HALT .. d	†		†		†	†		†	†		†		†		†	†		†	†		†	†	
STANLEY BRIDGE HALT d	7 14	..	8 35	..	9 24	11 19	..	13 19	14 14	..	15 23	..	16 15	..	16 59	..	..	18 30	19 20	..	20 04	..
CHIPPENHAM .. a	7 20	..	8 41	..	9 30	11 25	..	13 25	14 20	..	15 23	..	16 15	..	17 05	17 45	..	18 30	19 20	..	20 10	21 48	..

Saturdays

20 June to 5 September

																A	A								
CALNE .. .. .. d	6 55	..	8 26	9 15	..	11 10	12 20	..	13 07	14 03	..	15 11	16 02	..	16 50	17 43	..	18 17	19 05	19 55	21 00	21 35	..
BLACK DOG HALT .. d	†	..	8 29		..	†	†	..	†	†	..	†	†	..	†	†	..	†	†	†	†	†	..
STANLEY BRIDGE HALT d	7 04	..	8 35	9 24	..	11 19	12 29	..	13 16		..			..	16 59	17 52	..		19 14	20 04	..	..
CHIPPENHAM .. a	7 10	..	8 41	9 30	..	11 25	12 35	..	13 22	14 16	..	15 23	16 15	..	17 05	17 58	..	18 30	19 20	20 20	10 21	13 21	48	..

Heavy figures indicate through carriages
For general notes see page 49

† Calls when required to set down and take up passengers. Those wishing to alight must inform the Guard at Chippenham or Calne, and passengers desiring to join should give the necessary hand signal to the driver. Trains depart from Black Dog Halt 3 minutes after leaving Stanley Bridge Halt and 3 minutes after leaving Calne. Trains not shown to call at Stanley Bridge Halt depart from Black Dog Halt 9 minutes after leaving Chippenham.

making their pilgrimage, so often acted out during those times on train services they were about to lose. I can't recall if we stopped at the intermediate station of Black Dog Halt, built because the line had crossed the Marquis of Landsdowne's estate and he insisted on having a station for loading his racehorses and other domestic items. It was, in effect, a private station, but locals were allowed to use it. These feudal arrangements gradually disappeared, with the station becoming a public facility in 1904 and elevated to normal status (i.e. in the public timetable) in 1953. Sustrans subsequently converted a major section of the track bed as part of the National Cycle Network.

I reflected during the near two-hour journey back to the smoke: another day, another trip, another line closure. Unrepeatable railway journeys, par for the course back then, to be documented in order that sometime in the future I might write about them for others to enjoy.

The massacre of WR steam continued unabated; indeed, I can't recall any parallels within the railway industry to rid itself of 'unwanted material' with such an indecent haste. With Westbury shed closing and its entire six-strong fleet of tanks being condemned at the end of September, the WR had a mere 162 steam locomotives remaining.

The iconic TV programme *Thunderbirds* was launched that month – with the Rolling Stones '(I Can't Get No) Satisfaction' and the Walker Brothers' 'Make it Easy on Yourself' vying for the top spot.

13

Demise of the Manors

Rather ironically, despite being the smallest (and in their early days the least successful) of Swindon's 4-6-0 classes, the thirty Manors proved to be the most resilient in the rundown of Western Region steam. Twenty were built between January 1938 and February 1939, with the balance of ten (delayed by the Second World War) being turned out by BR in November and December 1950. They were the last GWR design to survive unscathed, the first withdrawal coming as late as April 1963. Their demise was as undignified as that of almost every other GWR engine: shamelessly run into the ground without name- or numberplates, much less the attention of a cleaner. With a further ten being condemned during 1964 and the balance the following year, 7808 and 7829 had the dubious accolade of being the final two at Gloucester lasting until that December. A remarkable nine (30 per cent of the class) have survived into preservation, albeit some as static exhibits (see Appendix IV).

There were so many places to go, so many classes of steam locomotives disappearing, and so many line closures, it wasn't until October 1965 that I returned to Shrewsbury, this being a knee-jerk reaction to the news regarding the imminent demise of the Manors. At the time, aware that they were still putting in sterling performances over the Cambrian line from there to the Welsh coast, it intrigued me as to the reasoning behind their withdrawal en masse. A letter to the editor of *Steam Days* magazine following an article of mine (November 2013) finally answered this query:

Demise of the Manors

'The Manors were condemned in November 1965 because the WR management wanted to turn Swindon works over to diesel repairs and no longer wanted the responsibility of maintaining ex GWR types.' The writer, Kevin Dakin, went on to state that he recalled reading in *Modern Railways* that trials had taken place with a Hymek and an EE Type 3, both borrowed from the WR, but nothing came of it and steam haulage continued using the inferior-powered but up-to-the-job BR Standard 4MTs.

The remoteness of the Cambrian system meant that for many enthusiasts the Manors were a greater challenge to cop, or in my case travel behind, than other GWR namers, being, by 1965, a rare sight at the locations they resided at elsewhere within the Western Region. The scarcity of services over the routes they did work did not necessarily entice haulage bashers such as myself because many hours were 'lost' when catching a run with one.

So let the journey begin and the early hours of Saturday, 16 October saw me arriving into Crewe, from London, at 0130 and, having located a warm compartment and snuggling down on the engineless 2150 York to Aberystwyth TPO/Mails – the 6ft-wide compartment seat making an ideal bed – I soon slipped into the world of nod. Although not departing Crewe until 0225, this train had an extended time there for the Post Office staff to work on. Awoken by the locomotive being detached at Shrewsbury, I just had enough time to run to the front and note Crewe South had turned out Carlisle Kingmoor-allocated 4-6-0 45295, in sparkling ex works condition, which had worked 1M41 the 32 miles from Crewe.

Although for most people, night-time is a time of relative quiet when they take to their beds and industry calms down, this wasn't the case on the railways. Back then, they were a twenty-four-hour affair and at 0400 that morning, Shrewsbury station was a hive of activity with newspapers, parcels and mail all being loaded into the Welsh departure.

Anyway, upon feeling the forwarding locomotive backing onto the train, I once again ventured out onto the platform and made my way past the many BRUTEs (British Rail Universal Trolley Equipment) to discover which Manor was to take me westward. It turned out to be 7802 *Bradley Manor*, which was smartly presented, albeit without brass cabside number-plate and nameplates. She was one of the remaining stud of eight examples at Shrewsbury and I stood for a while, the smell of coal filling the air, watching the crew preparing their machine for the arduous journey ahead. It was as if she was looking forward to the challenge with her safety valve preparing to 'pop', the roaring fire being fed and the heat generated off her warming all around on a cold, dark morning – as living and breathing a machine as ever there was. It was as if she were speaking to her crew, with the boiler filled to its maximum and bubbling away together with various groans of her metal expanding, reflecting the increasing heat being generated.

After performing a shunt move (with me aboard) to attach a van in the bay platform, we departed bang on time at 0410. In the darkness of an early morning, those wonderful sounds emanating from *Bradley Manor* as she barked through the Welsh hills over the 81 miles on undulating gradients will stay in my memory forever.

A small independent company, the Shrewsbury & Welshpool Railway opened this line between the two locations in 1861 with the idea being to connect with the Cambrian Railway from Oswestry. The 23-mile-long line west to Machynlleth was opened two years later and, because it followed the natural contours of the land, it was, in comparison to many, economical to construct. This led to a rather meandering line with the only major construction sites being the twin-bridge crossing of the Severn at Penstrowed and the 120ft-deep cutting at Talerddig, at the time the deepest railway cutting in the world, in order to cross the Cambrian Mountains.

Having arrived into Aberystwyth at just gone 0700, *Bradley Manor* had a two-and-a-half-hour turnaround prior to working back to Shrewsbury with the Cambrian Coast Express. Although the sun was beginning to rise, it was still quite chilly and, anticipating many fruitless hours festering there, I purchased a paper and returned to the warmth of the station waiting room. Looking back, I am unsure why I leafed idly through the timetable, but on doing so I realised that, by catching the 0740 DMU to Dovey Junction, I could venture up the Welsh coastline as far as Barmouth. This would enable me to intercept the Pwllheli portion of the Cambrian Coast Express, thus reaping an unexpected additional run with another steam loco. I didn't, however, run the risk of a cross-platform connection at Barmouth, instead alighting a station short at Morfa Mawddach. There, Machynlleth's Standard 4MT 75002, one of the dozen BR 4MTs destined to take over the Manors' workings for the remaining seventeen months of steam haulage over the Cambrian, was that day's Pwllheli portion power. This 'fill-in' neatly completed my coverage of the coast-clinging line alongside Cardigan Bay, a line sufficiently scenic to describe here.

Saturday, 16 October 1965 sees Machynlleth-allocated, 14-year-old, Swindon-built BR 4MT 75002 arriving at Morfa Mawddach (née Barmouth Junction) with the Pwllheli portion of the up Cambrian Coast Express.

Departing Morfa Mawddach, we passed the wide sweep of Fairbourne Bay before diving through an avalanche shelter squeezed between the 2,930ft Cader Idris and the beach, clinging to the cliffs on a tightly curved 15mph speed-restricted narrow ledge. After descending alongside Friog Cliffs, we called at Llwyngwril before making our way through a maze of sand dunes to Tywyn and then crawling over the rooftops of Penhelig. Finally, having passed alongside the Dovey estuary, we arrived at the isolated Dovey Junction station.

At the road-less, wooden platformed Dovey Junction, *Bradley Manor*, having arrived with the Aberystwyth portion earlier, took over the entire train and, unlike the journey earlier that day, I was able to enjoy the vista awarded to travellers through central Wales, starting with an initial gentle climb to the line's summit at Talerddig. After the closed station of Commins Coch Halt was passed, the climb steepened to 1 in 60 before a short level section through the similarly closed Llanbrynmair station. The gradient then increased with 3½ miles of 1 in 52/56 – easing slightly shortly before the summit. A healthy

Sister 75004 being passed at Llwyngwril while working a northbound freight. This then Machynlleth-allocated locomotive was transferred to Shrewsbury upon 6F's closure to steam in December 1966, only to succumb to withdrawal when 6D closed to steam in March 1967.

noise was emanating from the Manor, which, together with her clutch of Shrewsbury sisters, had mastered this terrain for many years. After rattling down through Carno and Caersws and calling at Newtown, we charged the bank up to Westbury before speeding downhill to Shrewsbury.

At Shrewsbury, with no steam services expected for several hours (that I knew of), I took the opportunity to visit the shed using the 'bunkers' Bible', Fuller's *Directory*, copping sixty steam locomotives and five diesels. On the scrap line, it was sobering to note the mix of former GWR and LMS locomotives, one-time rivals from competing companies, clustered together – rusting, unwanted, ignored, but sitting together in one final handshake awaiting their inevitable death sentence:

> The shed is on the east side of the Shrewsbury–Hereford line south of the station. Turn left out of the station along Castle Gates and continue into Castle Street. Turn left into St Mary's Street, continue along Dogpole and turn left into Wyle Cop. Cross the English Bridge and turn right along Coleham Head. Fork left along Betton Street and turn left along Scott Street (over two railway bridges). The shed entrance is a door on the left-hand side. Walking time 25 minutes.

Two views of 27-year-old Collett-designed and now preserved 4-6-0 7802 *Bradley Manor*. The first shows her departing Machynlleth with the up Cambrian Coast Express and the second at rest on Shrewsbury shed after her exertions.

16/10/65	6D Shrewsbury @ 1400
In steam	6947 *Helmingham Hall* (81F), <u>7802 *Bradley Manor*</u>, <u>7820</u> *Dinmore Manor*, <u>7822 *Foxcote Manor*</u>, 9657, 44821, 44936 (2D), 45031 (6A), 45078 (8B), 45132, 45198 (6C), 45250 (6A), 45348, 46518 (8C), 48440 (6C), 73025, 73040 (6C), 75038, 75053, 78058, 84000 (6C), 84004 (6C)
Dead	3709, <u>7819 *Hinton Manor*</u>, 7821 *Ditcheat Manor*, 44814, 45052 (2A), 45285, 45311, 45419 (6A), 46508 (6C), 48404, 48418, 75016, 78035, 78038, 78060
Withdrawn	3817 (6C), <u>3850</u> (6C), <u>3855</u> (6C), 6604 (6C), 7801 *Anthony Manor*, <u>7827 *Lydham Manor*</u>, <u>7828 *Odney Manor*</u>, 41209, 46510, 46511, 73036, 73038, 73090, 73167, 80048, <u>80072</u>, <u>80078</u> (6C), <u>80079</u> (6C), <u>80080</u> (6C), <u>80100</u>, 80101 (6F), <u>80135</u>, <u>80136</u>
Diesels	D1635, D1842, D3193, D3194, D3970

Note: Underlinings indicate the fifteen locomotives that survived into preservation.

Croes Newydd-allocated 2-8-0 3855, withdrawn two months earlier, awaits the call of the cutter's torch. She was, however, one of the fifteen locomotives seen on my visit to Shrewsbury shed that survived into preservation. Having spent twenty-two years at Barry, she is now at the East Lancashire Railway.

Lifelong Shrewsbury resident 19-year-old 0-6-0PT 9657 is in light steam. Visiting Croes Newydd's 2-8-0 8F 48440 lurks in the background.

Delivered new to Oswestry in 1952, upon that shed's closure in January 1965, Swindon-built Ivatt 2MT 46511 was dispatched to 6D, where she is seen rusting away.

With little revenue-earning usage for a 2MT locomotive, Croes Newydd's 2-6-2T 84000 (with sacking over its chimney to stop birds nesting) lies dead at the back of Shrewsbury shed.

On the return walk, I saw 7822 *Foxcote Manor* (a locomotive I finally caught up with fifty-two years later) passing Sutton Bridge signal box with the down Cambrian Coast Express. Noting there was nothing allocated on the shed's board for the 1531 for Birkenhead Woodside, I was relieved to see, upon returning to Shrewsbury station, green-liveried Croes Newydd-allocated 73040 arrive for the train; one of the two, the other being 73025, Standard 5MTs was seen in light steam on 6D earlier.

Opened in 1848, Shrewsbury station, partially built over the River Severn, was once a rich tapestry of steam with both GWR and LMS types making regular forays to the area. After the 1963 boundary changes, the majority of GWR types (except the Manors for the Cambrian services) had deserted the area, leaving just the BR and LMS types for me to witness. With the station awarded Grade II status in 1969, on platform 3 there is a plaque listing the forty-two local railway employees who lost their lives in the First World War. With nowadays three through roads and two south-facing bays, the former platforms 1 and 2 (from where the 0410 Aberystwyth departure went) have long since fallen into disuse. An eighth platform hidden behind a wall was allegedly once used to transport prisoners to and from Shrewsbury gaol. The

Shrewsbury's 7822 *Foxcote Manor* storms past Sutton Bridge Junction with the down Cambrian Coast Express. I was en route back to Shrewsbury station having bunked 6D and it was to be fifty-two years before I was able to catch a run with her (see p. 203).

impressive 1904-built Severn Bridge Junction signal box can be seen from the south of the station. At 38ft tall, this 180-lever (of which only 50 per cent remain in use) iconic structure is a listed building and is now the largest surviving mechanical signal box left in the world.

I then took the 1531 Birkenhead Woodside departure over, to me, uncharted territory. Nowadays marketed as the Severn-Dee line, this 42½-mile line was to become (as fully described in Chapters 17 and 19) a racing track in the final days of steam. On this occasion, I sat back and enjoyed the sound of a seemingly fit Standard 5MT bellowing her way through the pleasant Welsh Marches. Having never viewed an aqueduct before, I marvelled at Telford's creation on the east side of the line as we crossed Chirk viaduct. Little was I to know that I would cross this in a narrowboat (feeling very queasy because of the sheer drop) some years later. It was, however, worth it, with a horse-drawn boat ride along the nearby Llangollen canal afterwards.

Back to the October visit and, after progressing along the speed-restricted (mining subsidence) Gresford bank, arrival into Chester station was another revelation as I was unaware that a reversal had to be made. Not alighting out of the train, it wasn't until we arrived at Birkenhead Woodside that I discovered the power over this final leg was a Fairburn tank. While at Chester I noted that Stanier 5MTs 44807 and 45004 were working Manchester Exchange services and a 2-6-4T, 42606, was also in the mix – causing me to write in my dog-eared notebook: 'This steam saturated area requires revisiting!'

The line from Chester to Birkenhead was opened by the Birkenhead Railway Company in 1840 and was absorbed into LNWR twenty years later. Always worked as a Joint Railway between LNWR and GWR until nationalisation, the line was privy, until March 1967, to six express trains to and from Paddington. With the line mostly being as flat as a pancake, the crews from both depots often reached some exhilarating speeds with the six-coach trains.

Perhaps just a couple of stations en route are worth a mention. First, there was Port Sunlight, opened to workers from the nearby Lever Brothers' factory in 1914 but not to the public until thirteen years later, with its overriding aroma of soap. Then there was Rock Ferry with its Mersey EMUs through the tunnel to Liverpool.

Taken from a passing train with dockside cranes in the background, Birkenhead's seemingly driverless 0-6-0T 47674 shunts in the yard opposite her home shed. Completed at Horwich in 1931, she was always a Wirral-based locomotive, at either Bidston or Birkenhead, remaining in service until December 1966.

Demise of the Manors

Just prior to arriving into Birkenhead, I witnessed the greatest gathering of BR 9F 2-10-0s I was ever to see. Aptly described by the late Robert Adley MP as 'a smoking crumbling relic of the Victorian era', the sulphurous haze over the surrounding area must surely have had a degenerating effect on the rows of washing on lines in the gardens of the nearby terraced housing. If Patricroft was the graveyard setting of the Standard 5MTs and Carlisle for the Britannias, then here, at Birkenhead's Mollington Street depot, was the same for the BR 9Fs. Upon nationalisation, locomotives of GWR, LMS and BR origin were allocated there. However, following the 1963 combination of Beeching cuts and regional boundary changes, the entire GWR fleet was dispatched to Swindon for reallocation or scrapping. Nevertheless, work on the heavy iron-ore trains between Bidston Docks and Shotton Steelworks was to extend the lives of the BR 9Fs for a further four years.

Woodside station, once the grand outpost of GWR, was opened as late as 1878 to replace the inadequate 1844-built Monks Ferry station. It displayed all the signs of neglect associated with uncared for steam-age infrastructure. Admittedly, it was 1730 on a cold and dark October evening, but first impressions reminded me of a terminus equivalent of Birmingham Snow Hill, with soot-encrusted metalwork and the feeling of desertion – the concourse having large, water-filled, uneven flagstones in spite of the supposed canopy. Bradshaw described it as 'a station of truly baronial proportion and worthy of any London terminal'.

Provided with five short platforms, the rails disappeared from the terminus sharply on a curve into a deep, vertical cutting and then, almost immediately, into a ½-mile-long tunnel beneath the town. It was an atmospheric, hemmed-in location, where the sound of the steam locomotives moving off resounded off the surrounding high walls. When the Paddington expresses were withdrawn in March 1967, the only services that remained were those operated with DMUs to Chester and Helsby, these being truncated at Rock Ferry upon Woodside's inevitable closure that November.

To complete that day's itinerary, I then made my way to Manchester for my first of six journeys on the GCR-routed 2250 Manchester Central to Marylebone (fully documented in my 2019 *Confessions of a Steam-Age Ferroequinologist*).

Time arr	dep	Station	Traction	Date
	2235	Euston	E3098	Fri, 15
0124	0225	Crewe	45295	Sat, 16
0317	0410	Shrewsbury	7802 *Bradley Manor*	
0711	0740	Aberystwyth	DMU	
0813	0820	Dovey Junction	DMU	
0915	0935	Morfa Mawddach	75002	
		Dovey Junction	7802 *Bradley Manor*	
1244	1531	Shrewsbury	73040	
		Chester General	42121	
1728	1740	Birkenhead Woodside	DMU	
1812	1920	Chester General	DMU	
1954	2023	Crewe	E3012	
2108	2127	Manchester Piccadilly	EMU	
2129		Manchester Oxford Road		
	2250	Manchester Central	42066	
		Guide Bridge	27005 *Minerva*	
		Sheffield Victoria	D6746	Sun, 17
		Leicester Central	44869	
		Woodford Halse	D5085	
0506		Marylebone		

The Manors' reign came to an end on 5 November. To replace them, a dozen BR 4MT 4-6-0 75xxxs were drafted into both Machynlleth and Shrewsbury sheds. Although there appeared to be several individual changes among the class, the numbers remained the same for the summer 1966 services. With Machynlleth's closure in December 1966, Shrewsbury inherited the best of them and at the cessation of Cambrian steam it had nine at its disposal, frequent substitutions of Croes Newydd examples being observed.

Elsewhere, both Ebbw Junction (Newport) and Llanelly sheds closed; the latter's eight tanks were condemned on the spot. Of the eleven locomotives allocated to Ebbw Junction, however, eight escaped, with two Granges going to Worcester, two Modified Halls to Oxford and two BR 9Fs to both Bristol Barrow Road and Gloucester.

Demise of the Manors

In other news that month, the headlines had been dominated by the Saddleworth Moors murders, with Myra Hindley and Ian Bradley being arrested and subsequently charged. On TV, *The Magic Roundabout* was first aired, while in the charts, the number one spot was monopolised by Ken Dodd's 'Tears'.

14

Oxfordshire Outings

At the end of October 1965, the WR had a mere 141 steam locomotives that were allegedly operational, located at eight sheds, to dispose of, comprising ninety-one GWR, nineteen LMS and thirty-one BR types.

With the commencement of the winter timetable, although the resourcing of the northbound York was shared between Bournemouth and Banbury, the southbound had become the sole preserve of a Banbury Black 5. And so, on a damp and foggy Thursday, 11 November, and with the chart-topping Rolling Stones' 'Get Off of My Cloud' not being entirely inappropriate, after travelling out of Waterloo with the subsequently preserved 73082 *Camelot*, I joined the northbound York, worked that day by sister 73115 *King Pellinore*. At Oxford, instead of the expected GWR steam a DL was turned out. However, even if it had have been steam I wouldn't have continued on it as the sparse returning train service would have 'stranded' me at Banbury for several hours.

In 1844, GWR opened the line I had travelled on that afternoon to a terminus at Grandpont in the southern suburbs of Oxford. Eventually extended to Banbury six years later, the Oxford station I was standing on was opened in 1852 and, to differentiate it from the adjacent LNWR 1851-built Rewley Street station, it was named Oxford General. At least the one-and-a-half-hour sojourn at Oxford wasn't boring, with several photographs being taken before returning south with Banbury's 44710 on the York.

SHED ALLOCATIONS OF BRITISH RAILWAYS LOCOMOTIVES

IN NUMERICAL ORDER

7 * 6F	4113 85A	6165 85A	6999 * 81F	9626 85A	31866 70C	34047 *70F
8 * 6F	4135 86C	6611 6C		9630 6C	31873 70E	34048 *70E
9 * 6F	4161 85A	6625 2A		9640 2B		34052 *70E
	4635 2A	6626 6C		9641 2C		34056 *70E
	4645 6C	6667 2A		9656 86E		34057 *70E
	4646 2C	6668 2A		9657 6D		34059 *70E
	4668 87F	6683 6C		9658 2B		34060 *70D
	4671 86E	6697 2D	7029 *85B	9669 6C		34064 *70E
1628 6C	4680 85A	6815 *86E	7002 *6D	9672 82E	30006 70A	34066 *70E
1638 6C	4689 85B	6819 *85A	7808 *85B	9680 82E	33020 70A	34071 *70D
1660 6C	4696 2C	6829 *85A	7812 *6D	9724 2C	33027 70A	34076 *70E
	4920 *82E	6831 *2B	7816 *85B	9773 81F		34077 *70D
		6833 *2B	7819 *6D	9774 2A		34079 *70D
		6838 *85A	7820 *6D	9776 2B		34082 *70D
		6847 *85A	7821 *6D	9789 81F		34086 *70D
		6849 *81F	7822 *6D			34087 *70D
		6853 *2A	7828 *85B		34001 *70A	34088 *70D
3605 2B	5605 2A	6855 *2A	7904 *82E		34002 *70A	34090 *70D
3607 2C	5606 2A	6856 *85A	7907 *82E		34004 *70F	34093 *70D
3619 2C	5658 2A	6859 *86E	7909 *81F	30064 70D	34005 *70E	34095 *70D
3625 2A	5659 6C	6861 *2A	7914 *82E	30067 70D	34006 *70E	34097 *70D
3675 85B	5676 6C	6871 *2B	7919 *81F	30069 70D	34008 *70D	34099 *70D
3677 81F	5677 6C	6872 *85A	7922 *81F	30071 70D	34009 *70D	34100 *70D
3681 82F	5971 *82E	6876 *85A	7924 *82E	30072 70C	34013 *70E	34101 *70D
3682 85A		6879 *2A	7925 *81F	30073 70D	34015 *70E	34102 *70D
3696 82E		6923 *81F	7927 *81F		34017 *70D	34104 *70D
3709 6D		6930 *2D			34019 *70D	34108 *70E
3744 2B		6932 *81F			34021 *70A	
3749 6C		6937 *81F		31405 *70C	34023 *70D	
3754 6D		6944 *86E		31408 70C	34024 *70F	35083 *70F
3758 82F	6106 81F	6947 *81F		31411 *70C	34025 *70F	35004 *70F
3759 85B	6111 81F	6951 *2D	8718 2C	31619 *70C	34032 *70E	35007 *70G
3775 85B	6113 85B	6952 *2D	8720 81E	31639 *70C	34033 *70D	35009 *70D
3776 2B	6114 86E	6953 *81F	8767 2B	31791 *70C	34034 *70D	35010 *70F
3782 2B	6126 81F	6956 *81F		31803 70C	34038 *70A	35011 *70F
3788 2B	6134 81F	6959 *81F		31809 70C	34040 *70F	35012 *70G
3789 6C	6135 81F	6960 *81F		31816 *70C	34041 *70D	35013 *70F
3790 88A	6136 81F	6967 *81F	9608 2C	31858 *70C	34044 *70F	35014 *70G
3792 2B	6141 82F	6984 *82E	9610 6C			
	6143 81F	6989 *85B	9613 2C			
	6145 81F	6990 *82E	9614 2C			
	6147 85A	6991 *81F				
	6156 81F	6993 *81F				
	6160 85B	6998 *81F				

	9626–41291
	35017 *70G
	35022 *70G
	35023 *70F
	35026 *70G
	35027 *70F
	35028 *70G
	35029 *70G
	35030 *70G
	41202 9B
	41204 9B
	41206 83G
	41207 10J
	41211 8K
	41212 15A
	41216 83G
	41217 12B
	41220 9B
	41222 12B
	41223 83G
	41224 70F
	41229 12B
	41230 70F
	41233 9B
	41234 6G
	41241 10G
	41244 8K
	41249 83G
	41251 10J
	41264 12B
	41272 10G
	41283 83G
	41284 70G
	41285 12B
	41286 8G
	41287 70D
	41290 83G
	41291 83G

Isle of Wight Locomotives

14 *70H	20 *70H	26 *70H	31 *70H
16 *70H	21 *70H	27 *70H	33 *70H
17 *70H	22 *70H	28 *70H	35 *70H
18 *70H	24 *70H	29 *70H	

The final *Loco Shed* book listings, as at 7 November 1965, of the ex GWR fleet.

Southall shed had fourteen examples of the 61xxs in July, but after four withdrawals and seven transfers to Oxford the following month, the remaining three were dispatched to Bristol Barrow Road, Gloucester and Worcester during that November; steam-servicing facilities for visiting LMR locomotives being retained to the end A not dissimilar scenario was acted out at Severn Tunnel Junction where, although steam-servicing facilities were kept to the year end, all bar two 9Fs were condemned. The remaining 9Fs were dispatched to Gloucester. Then it was Bristol Barrow Road's turn, its entire sixteen-locomotive allocation, including the subsequently preserved 4920 *Dumbleton Hall*, being dragged away to Dai Woodham's scrapyard on Barry Island.

With 5042 *Winchester Castle*, 7022 *Hereford Castle* and 7034 *Ince Castle* having been withdrawn that June, it was left to 7029 *Clun Castle* to fly the flag for the class. She was sporadically turned out for the 1700 Gloucester to Cheltenham as well as working the final steam departure out of Paddington, a BR-sponsored rail tour to Bristol and Gloucester on 27 November.

Having just crossed over the Oxford Union Canal, 22-year-old home-allocated 6953 *Leighton Hall* shuffles past a PW gang (not a HV vest in sight!) at Oxford with a short freight on a very cold day. Note the frost on the wooden sleepers. The run down of WR steam is epitomised by this Thursday, 11 November 1965 view: coated in filth, devoid of all identification, steam leaking from every orifice but still doing the job it was designed for. Once hauling prestige passenger trains, the neglect and uncared for external condition, with her name- and numberplates missing, was how the WR authorities, having not filled engine cleaner vacancies for some time, treated a once fine example of British workmanship. Now just days away from the cutter's torch, it made for a sorrowful sight.

Another Oxford resident, Standard 5MT 73166, scuttles past the Pines Express at Oxford station.

One of the many 61xx 2-6-2Ts to gravitate to Oxford in the final months of WR steam was 24-year-old Collett-designed 4MT 6134, seen shunting Rewley Yard, having arrived from Southall in January 1965. Seventy of these Collett locos were built, at least twenty-six of them spending time on the books at Oxford in the 1960s and all bar one of them scrapyard-bound when their use ended in December 1965.

Transferred in from Didcot in May 1964, 2-6-2T 6126 runs light engine past Oxford's platform, which is jam-packed with the long-lost parcels traffic.

The accolade of being the last steam arrivals into Paddington, however, went to two Stanier 5MTs, which arrived 'inside' two DL-hauled trains from Birmingham on 22 January 1966 having provided the steam heating.

So now, on 1 December, excluding the two continuing S&D locomotive depots at Bath and Templecombe, just three WR depots retained steam allocations: Oxford, Gloucester and Worcester. With other sheds having dispatched their best steam locomotives to them, whether they wanted them or not, the eighty-eight locomotives shared between these sheds would have probably found little use. Oxford, for example, had, after the summer Saturday services had finished, condemned a good many passenger locomotives, only to receive a deluge from closing sheds elsewhere! Weeks later, Worcester shed closed to steam, with its allocation of Panniers, 61xxs and BR 5MTs all being condemned.

After briefly visiting Reading West on a SR-based outing on Saturday, 11 December and reaping a run with another Banbury resident – namely 45299 – I made my final visit to the Cherwell Valley steam scene on a very cold and frosty Tuesday, 28 December. Having travelled down on the 0930 out of Waterloo with Standard 5MT 73081 *Excalibur*, I collected a run with filthy WC 34040 *Crewkerne* on the northbound Pines Express.

A derailment near Reading West was delaying all through services and with a thirty-seven-minute late arrival into Banbury, I thought I wouldn't have so much waiting time in the cold. How wrong I was! The York train, having arrived a mere twenty minutes late, suffered a further thirty-minute delay due to a broken train heating pipe having to be replaced. During this extended wait, 2D's 45331 arrived annoyingly with the northbound York. Times were a-changing because the station pilot that day, rather than a GWR locomotive, was Brit 70053 *Moray Firth*. Although I appreciated the required haulage of the subsequently preserved BR 4MT 75069 on the York, she obviously wasn't a well bunny and achieved a maximum of a mere 58mph en route, with the eleven-coach, one-van train depositing me ninety minutes late at Basingstoke.

In other news, the 70mph motorway speed restriction came into force on the 22nd and in the charts (there were never any Christmas-orientated hits back then – whatever was at the top stayed there) the final chart topper of the year was the double-sided Beatles hit 'Day Tripper'/ 'We Can Work it Out'.

The very last Hall to be built, in February 1949, was Modified Hall 6999 *Capel Dewi Hall*. Seen here at Didcot in an appalling external condition, on Tuesday, 28 December 1965, she had the 'honour' of being Oxford station's last steam pilot on Monday, 3 January 1966.

Just days away from her demise, Guildford-allocated N 2-6-0 31809 is seen at Didcot North Junction when passing by on the WC 34040 *Crewkerne*-hauled Pines Express.

On yet another cold and frosty day, Banbury's 9F 92218 creeps up to a signal north of the station. Transferred to Speke Junction, she became the final 9F that worked a passenger train, albeit a railtour, with me aboard, in April 1968, just days away from withdrawal.

Minutes later and Banbury's 7-year-old 92247 trundles through with a 'lightweight' freight.

Unlike a lot of her sisters, 2-6-2T 6111 was a long-term Oxford resident. This was my last photograph, taken from the 75069-hauled York–Bournemouth train, of an active Western Region-allocated steam locomotive. She is shunting at the former LNWR Rewley Street station.

WR steam was officially scheduled to end on 31 December 1965 but because of the vagaries of the calendar, the actual date was Monday, 3 January 1966. On 1 January, perhaps as an 'up yours' to Paddington management, *Clun Castle* was once again turned out for the 1700 out of Gloucester. Also on the 1st, 7924 *Thorneycroft Hall* worked the Banbury to Basingstoke perishables forward from Oxford, returning light engine. On the bank holiday, the 3rd, a specially cleaned 6998, with wooden replica *Burton Agnes Hall* nameplates, smokebox straps painted white and cabside but not smokebox numberplates returned, worked the Oxford to Banbury leg of the 1025 Poole to York, 1410 off Oxford. She had undergone a heavy general repair at Swindon in January 1964 and was the best of Oxford's remaining GWR 5MTs, conveying the Lord Mayor of Oxford on the footplate from shed to station. Running light engine back to 81F, she was then placed in store, the Great Western Society having purchased her for £2,500. With the grapevine not nearly as efficient as today's due to a lack of mobile phones and internet, I was unaware of this bank holiday running. Had I known, I would have moved heaven and earth to be there! Oxford shed had become a dumping ground en route to the South Wales scrapyards, Pannier 9773 being steamed on the Monday to sort them out.

Thirty former GWR locomotives remained within the 84-coded group on the LMR at Tyseley, Oxley, Stourbridge, Shrewsbury and Croes Newydd, comprising three 16xxs, twenty-five 57xxs and two 56xxs; Panniers 4646, 4696 and 9774 were reported to have been still putting in a day's work until their withdrawal in October 1966.

My year-end total of GWR locomotives I was hauled by was a measly eleven, which if added to the 1964 total of ten, accrued to a wretched twenty-one. Thank goodness for the subsequent Barry scrapyard phenomenon, from which 130-plus former GWR locomotives were extracted, and the wonderful work carried out by hundreds of volunteers in today's preservation movement. As a result of their efforts, my running total is now an amazing eighty-four.

Former GWR steam locomotives 'in service' on the Western Region on 31 December 1965

Shed	Locomotive
81F	3677, 4920, 5971, 6106, 6111, 6126, 6134, 6135, 6136, 6141, 6145, 6147, 6156, 6160, 6165, 6849, 6923, 6932, 6937, 6953, 6956, 6959, 6967, 6984, 6990, 6991, 6993, 6998, 7904, 7907, 7914, 7919, 7922, 7924, 7925, 7927, 9773, 9789
82F	3681
85A	4680, 6847, 6951, 6952, 9626
85B	3675, 3758, 3775, 4689, 6848, 7029, 7808, 7829, 9672, 9680

15

Mission Accomplished

On 1 January 1966 3,003 steam locos, a decrease of 1,987 during 1965, remained on BR's books. A reduction of a further 1,305 was to occur during 1966.

On and from Monday, 3 January, the WR retained a grand total of twenty-one steam locomotives – whether they were all operational remains a moot point – to work their commitment to the remaining freight and 'sulky' passenger services over the S&D route. They comprised two 57xxs, two Jinties and four LMS 8Fs at Bath Green Park, with nine Ivatt 2MTs and four BR 80xxx 4MTs at Templecombe. Together with usage of Bournemouth's Standard Moguls and 4MT tanks, they were all that was required.

The S&D was to finally die on 7 March 1966, the appalling sulky service soldiering on for the two months before the substitute bus service supplier got its act together. The WR therefore, belatedly, became the first BR region to dispense with steam: mission accomplished. The WR had been selected by the BRB as being the first to do away with steam as there were no electrification plans for the region. Was there a phone call made that morning from the BRB to the WR headquarters at Paddington, congratulating them on achieving their objective? Promised bonuses within their next pay packet perhaps? The irrational haste to replace steam with a series of short-lived, unreliable diesel locomotives, at a massive cost to the taxpayer, was conveniently overlooked. For sure, steam locomotives from other regions still darkened WR territory, but they were just passing through. There were now no steam locomotives on the WR's books.

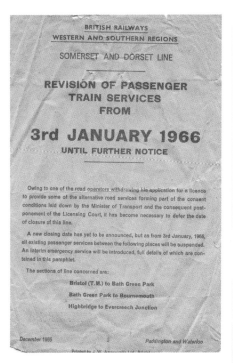

The S&D 'sulky' service on offer during the first eight weeks of 1966.

Mission Accomplished

The delayed closure therefore allowed one further *final* visit to the line. On the Saturday prior to closure, 26 February, and having once again travelled via Eastleigh and the Hymek DL train from there to Bristol, Alan and I were to be seen alighting at the Temple Meads station just before 0500.

Deliberately, for whatever reason, not advertised in the public time-table pamphlet as a through service, our 0600 departure for Branksome (Bournemouth West having closed in October 1965 and the SR authorities not being predisposed to deal with the majority of these inconsequential services at the Central) had sufficient dwell time at Bath Green Park to allow a visit to the shed. With just four daily trains across the Mendips, two on the Highbridge branch and five south of Templecombe, it really was not a customer friendly service on offer.

Leaving Bristol behind Warship D826 *Jupiter*, the first call was at the once-triangular platformed Mangotsfield station. It was here, having on one occasion been stranded overnight, that the *Dad's Army* actor Arnold Ridley (Godfrey in the series) penned the play *The Ghost Train*.

26/2/66	82F Bath Green Park @ 0630
In steam	3681
Dead	41291 (83G), 80039 (83G)
Withdrawn	48444, 75072 (83G), 75073 (83G), 92243 (82E)

Having visited 82F for one final time, upon returning to Green Park station we found sorry-looking, externally unkempt Bournemouth Mogul 76011 at the head of our train. Unlike my last visit the previous September, I was *not* inclined to window hang en route – not only was it still dark but bitterly cold to boot, with a layer of snow all around. What we witnessed was a sorry sight to behold: the gas-lit empty stations, the customer less trains, station staff wondering what their future was to be. Dripping with damp nostalgia, the semaphore signals still safely saw us on our way over the Mendips.

All this was observed from the wonderful warmth created by the steam heating. With dawn finally breaking at about 0700, a few photographs were taken by poking the camera out of the window of the compartment. I did, however, alight briefly at Evercreech Junction to snatch a photograph of the Mogul sating her thirst. As if to memorialise the line, a nearby public house was sympathetically renamed The Silent Whistle several years later.

Taken from the 76011-hauled 0600 Bristol Temple Meads to Branksome train on Saturday, 26 February 1966, this is the 55ft-high Midford viaduct over which the double-tracked section north of the adjacent station became single.

Shoscombe and Single Hill Halt, the last station to be opened on the S&D in 1929.

Nowadays wonderfully restored by the Somerset & Dorset Heritage Centre, what did the future hold for the signalman looking out of his box a week before the line closed?

At Shepton Mallet Charlton Road, Brighton-built Standard 4MT 80043 is working the 0700 Templecombe to Bath Green Park.

Bournemouth-allocated, 13-year-old BR 4MT 76011 calls at Evercreech Junction. This Mogul, although looking in an absolutely deplorable external condition, was obviously mechanically fit enough to last until the very end of SR steam in July 1967.

Taken from the train, Ivatt 2MT 41296, one of the eight WR-allocated locomotives retained to resource the final timetable, stands forlorn at Templecombe shed and would probably never turn a revenue-earning wheel again.

After the unnecessarily lengthy thirty-six-minute call at Templecombe, we were led down the bank from the 'Upper' by the depot's Ivatt 2MT 2-6-2T 41290, eventually arriving into Poole some four and a half hours after leaving Bristol – a mere 82¼ miles away! The only other active S&D locomotives noted that morning were LMS 3F 47506 at Radstock, Ivatt 41307 on the Highbridge branch and 80043 on the 0700 Templecombe to Bath. Bournemouth's 76026 and 80013 were on the 0705 and 0937 Bournemouth to Templecombe trains respectively.

What we witnessed that day will stay with us forever. A once splendidly run system was now just days away from closure, and neglect, decay and abandonment observed everywhere. With our mood lightened upon entering the SR steam scene, my travelling companion Alan continued onto the Waterloo services for the remainder of that day, while I obtained new steam track to Brighton on the once-daily Plymouth.

Two days after our visit, the Labour Prime Minister Harold Wilson called a snap election in the hope of increasing his vulnerable single-seat majority in Parliament, while earlier that month the Action Man toy was launched in the UK. Chart toppers were the Overlanders' 'Michelle' and Nancy Sinatra's 'These Boots are Made for Walkin''.

Mickey 41290 leads the 0705 from Bournemouth Central up the incline from Templecombe Lower.

With the WR wanting nothing more to do with steam, the responsibility of the motive power provision for the York train (the Pines Express having gone DL-hauled throughout) fell to Banbury, although it has to be said there were frequent observations of Bulleid Pacifics on it. Only water replenishment stops – i.e. no crew changes – were allowed at Oxford.

Seventeen Stanier 5MTs remained on Banbury's books at the beginning of 1966 but within weeks, with Colwick (Nottingham) taking over resource provision for the GCR services, five were dispatched there. Of the remaining twelve, I caught five of them on SR metals while working the York train, an example catch being Saturday, 9 April, with 44872 on the northbound and 45493 on the southbound. The gradually decreasing numbers culminated in 2D's closure that October.

One further visit to Banbury was made a week before the GCR closed, on Friday, 26 August that year. I had travelled out of Paddington on that evening's 1810 departure to connect into the 1905 Swindon to Sheffield Victoria. This 2116 departure out of Banbury was worked by Hymek D7009 as far as Leicester Central, where Tyseley's 4-6-0 44865 took over for the 61-mile journey to Sheffield, the purpose of my journey being 'required' steam track north of Nottingham. The final York train on Saturday, 3 September was worked by 34034 *Honiton* on the northbound and 34005 *Barnstaple* on the southbound, the train being diverted via Birmingham and DL-hauled throughout from the following Monday.

16

Rail Tour Bonanza

With increased monies from promotion within the clerical ranks of BR, I could now afford to travel on rail tours. The first of my five undertaken during 1966 and involving WR territory was on the A4 Commemorative Rail Tour, which used visiting St Margarets' A4 60024 *Kingfisher*, on Sunday, 27 March. This was the same day that Pickles the dog famously found the stolen Jules Rimet trophy in a south London garden four months before the World Cup football tournament kicked off in England.

Starting out of Waterloo at 1005, it was to be a straightforward out and back to Exeter via the former L&SWR route. Plagued by speed restrictions as far as Basingstoke Nine Elms, Driver Porter (who else!) managed to get 86mph at West Byfleet and 84mph out of the streak approaching Andover. After a crew change at Salisbury, several further 80mph-plus speeds were notched up prior to single-line working delays in the Honiton area that caused a sixteen-minute late arrival into Exeter St Davids. The day really was a pleasurable 345-mile experience, with sustained high-speed running on the return attaining numerous low 80mphs and meaning we arrived into Waterloo just two minutes late at 1937.

Moving on to Sunday, 14 August, and the A2 Commemorative Tour, utilising Dundee's 60532 *Blue Peter*, ran from Waterloo via the L&SWR to Exeter, returning up the GWML to Westbury and then via Salisbury to Waterloo. The Scottish locomotive was, however, not a well bunny and, having stalled surmounting Honiton bank, upon arriving into Exeter it was sent to Exmouth

2nd · SPECIAL EXCURSION
The Locomotive Club of Great Britain
A4 COMMEMORATIVE RAIL TOUR
(C.M. 1935) Sun. 27th. MAR. 1966

WATERLOO-BASINGSTOKE-SALISBURY-
YEOVIL JC. · EXETER CTL. · EXETER ST.
DAVIDS-EXETER CTL. · YEOVIL JCT.
SALISBURY-BASINGSTOKE-WATERLOO-
(S) FOR CONDITIONS SEE OVER

LNER 30-year-old A4 60024 *Kingfisher* poses at Exeter St Davids on Sunday, 27 March 1966 while working the A4 Commemorative Rail Tour.

Junction for examination. After attention at 83D, the tour was now running 162 minutes late and upon arriving into Westbury the decision was made that the booked locomotive change to Brit 4 *William Shakespeare* would take the train through to Waterloo (rather than the A2 regaining the train at Salisbury), running via Andover, rather than Fareham and the Pompey Direct, meaning it arrived into London just eighty minutes late.

The privately sponsored rail tour, Steam Again to the West Country, ran on Saturday, 15 October, with all profits going to the SR orphanage at Woking. It was a straightforward circular tour from Waterloo (departure at 0909) which, having traversed the GWML between Westbury and Exeter St Davids, was then assisted by Warship D866 *Zebra* up the 1 in 36 to Exeter Central, before returning via the former L&SWR main line. This eight-vehicle tour certainly lived up to its billing with superb performances by all the locomotives and footplate staff involved.

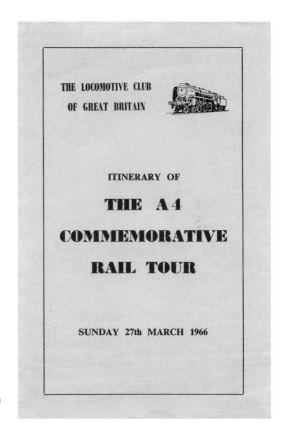

THE LOCOMOTIVE CLUB
OF GREAT BRITAIN

ITINERARY OF

THE A4
COMMEMORATIVE
RAIL TOUR

SUNDAY 27th MARCH 1966

Below: Included in this book because Mortimer was a WR station, Maunsell Moguls 31639 and 31411 pause for a photographic stop on Sunday, 3 April 1966 while working the Locomotive Club of Great Britain (LCGB) Wilts & Hants Rail Tour.

Axminster and an ailing A2 Pacific 60532 *Blue Peter* calls for water on Sunday, 14 August 1966 while working the A2 Commemorative Tour.

Prior to *Oliver Cromwell's* stranglehold on rail tours, Stockport's 70004 *William Shakespeare* was the favoured Brit. At Westbury she is seen taking over the A2 Commemorative Rail Tour, working it through to Waterloo instead of the intended Salisbury changeover.

The first leg from Waterloo to Westbury was worked by 8P 35023 *Holland-Afrika Line*, which, with Nine Elms Driver Bert Hooker, was also plagued by speed restrictions as far as Basingstoke. Thereafter, however, following prolonged running of 90mph, maximised at Andover Junction, a breathtaking 102mph was enjoyed. From Westbury, sister Merchant Navy 35026 *Lamport & Holt Line* was now in charge and a schedule of seventy-five minutes had been requested of the WR between there and Exeter but was refused – the eventual ninety-one minutes being bettered by, you've guessed it, sixteen minutes, with Whiteball summit being mounted at 55mph. More high-speed running was enjoyed up the former L&SWR route with a maximum of 88mph approaching Templecombe.

Finally, with Merchant 23 retaking the train from Salisbury, two further 90mphs were achieved at Andover and Hook before a 91mph approaching Brookwood. After arriving into Waterloo at 1719, was I satisfied with more than 350 miles of high-speed steam that day? No sir, I topped it up with a further 95½ miles on a Basingstoke return trip before wending my way home.

Saturday, 12 November, and the reason for my attendance on the Shakespearian Rail Tour was primarily the usage of the recently preserved Castle 7029 *Clun Castle*. Departing out of Waterloo at 0832, the legendary Nine Elms Driver Porter attained a seat-grabbing 75mph through Bracknell with the nine-vehicle train on WC 4-6-2 34015 *Exmouth*, the requested loco, 34102 *Lapford*, having failed earlier that week. With WR metals having been attained at Reading, MN 35023 *Holland-Afrika Line* took us on the 50-mile leg to Banbury, where the somewhat grimy *Clun Castle* took over for the run to Stratford-upon-Avon and maxing at 74mph down Hatton bank.

The appropriate usage of Britannia 70004 *William Shakespeare* then took the tour the 37 miles (during which her brick arch collapsed!) via Birmingham Snow Hill to Stourbridge Junction, where the planned usage of two GWR pannier tanks along the Town branch was replaced by a DMU 'due to the shortage of fit locos'. Retracing our steps to Birmingham (the

It's Saturday, 12 November 1966 and the Shakespearian Rail Tour calls at Reading General to change locomotives from WC Light Pacific 34015 *Exmouth* to MN 8P 35023 *Holland-Afrika Line*. Although only working the tour the 50 miles to Banbury, she also returned it the 72 miles back to Marylebone that evening.

The now privately owned 7029 *Clun Castle* at Banbury while participating in the Shakespearian Rail Tour. Having worked the final booked steam train out of Paddington in June 1965, she was withdrawn at Gloucester that December and purchased for £2,400 (scrap value) the following year.

planned route via Dudley and Swan Village not being accessible because of route availability restrictions and poor track condition), *Clun Castle* stopped additionally at Snow Hill for assistance, in the form of Chester's 45250, for the 1½ miles to Bordesley Junction, after which we were twenty-two minutes down upon arrival into Banbury.

The Merchant was waiting for us there to take us the 72 miles to London, but a further delay at Old Oak Common awaiting a pilotman, with the train being routed via Kensington Olympia, led to a fifty-five-minute late arrival into Victoria at 1855.

Another day – another rail tour! The West Country Special the following day, Sunday, 13 November, was the *very last* steam train to the West Country. The specially requested Brunswick green-liveried Standard 5MT 73029 had failed earlier that week with injector problems, her respectable sister 73065 acting as an adequate stand-in. Departing at 0958 out of Victoria's platform 7, the train took a circuitous route through the SR's then Central Division and Redhill, gaining WR territory at Reading.

Somewhat appropriately, Brit 4 *William Shakespeare* prepares to work the 37-mile leg of the Shakespearian Rail Tour from the bard's former hometown of Stratford-upon-Avon to Stourbridge Junction.

The planned usage of the tour train being top and tailed with two pannier tanks along the Stourbridge Town branch failed to materialise due to the non-availability of any surviving members of operational use. The substituted DMU is seen here.

The now Tyseley-based *Clun Castle* at Stourbridge Junction about to return the tour to Banbury. Also shown opposite is the ticket, brochure and map.

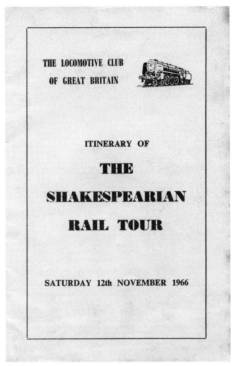

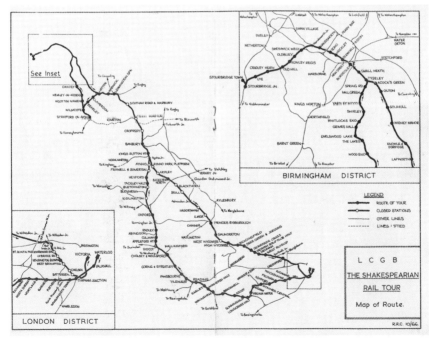

Sunday, 13 November 1966 and the West Country Special Rail Tour changes locomotives from Standard 5MT 73065 to WC Unmodified 34019 *Bideford*. She distinguished herself on the return up the ex L&SWR route, maxing at 86½mph with the seven-vehicle train on the downhill section east of Crewkerne. Also shown below is the brochure and ticket.

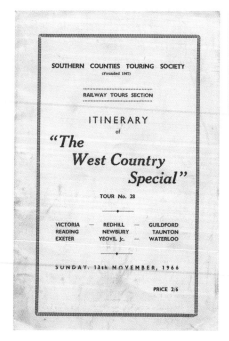

Arriving into Westbury a few minutes late, the Standard was replaced by a relatively clean WC Light Pacific 34019 *Bideford* for the run down the GWML to Exeter St Davids. Having been assisted up the bank to Exeter Central by Warship D818 *Glory*, the Bulleid really took off, covering the 48 miles to Yeovil Junction in forty-nine minutes with speeds of 79½mph down Honiton bank and 86½mph approaching Yeovil Junction. That performance thoroughly resonated with the Beach Boys' 'Good Vibrations', which was holding the top spot that week. Running just twelve minutes late, an engine change to MN 35023 *Holland-Afrika Line* (affectionately known for reasons undetermined to SR enthusiasts as 'The Pram') was made and, with five instances of 80mph-plus en route, we arrived into Waterloo just two minutes late at 1942.

Westbury sees another locomotive change, this one from BoB 34057 *Biggin Hill* to WC 34013 *Okehampton* on the Bridport Belle Rail Tour on Sunday, 22 January 1967. The train was already an hour late due to a body having to be removed from the train at Basingstoke; the much greater delays encountered on the Bridport branch have been adequately detailed in *The Great Steam Chase* (2013).

17

The Chester Chronicles

WARNING – The remainder of this book contains a proliferation of locomotive numbers. While in previous tomes I have steered away from documenting such minutiae, here I consider it essential in order for the reader to appreciate the addictive nature of a haulage basher. Why else would someone return time and time again to an area dominated by the ever-present Stanier 5MTs if it wasn't for the obsessional trait to catch runs with as many different examples as possible?

Sadly, being so late on the scene, I missed out on witnessing and travelling on the Paddington express services when operated by former GWR steam locomotives. Regional boundary changes in 1963 had transferred the ex GWR lines north of Banbury to the LMR and so by the time of my visit the services were operated by WR-allocated Brush Type 4s south of Shrewsbury and, with the exception of some Cambrian line services for which a handful of Manors had been initially retained, LMS and BR steam locomotives to the north. Indeed, the LMR lost no time in running down the ex GWR locomotives at its inherited depots, particularly Shrewsbury and Tyseley, and replacing them with the aforementioned types. Trains were provided with buffet/restaurant-car facilities as far as Wolverhampton, from where reduced formations progressed either to Birkenhead (with a reversal at Chester) or Aberystwyth.

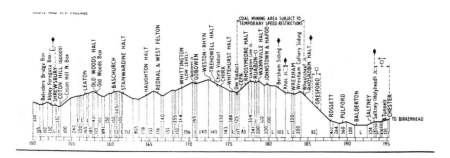

Gradient profile of the Shrewsbury to Chester line.

Perhaps here it is opportune to describe the route between Shrewsbury and Chester over which I was to spend a considerable amount of time, particularly during the final four weeks (February–March 1967) of steam along the line. With several fairly steep gradients and many level crossings, despite the short formation of most trains, the route was not an easy one to work and the speed restrictions north of Wrexham were not conducive to maintaining the tight schedules.

At the time of my travels, Shrewsbury station suffered at peak times because of its limited platform accommodation, exacerbated by nearly every train from Paddington either changing locomotives or, if destined for the Cambrian Coast, reversals. The early 1960s rebuild, during which it lost its train shed canopy, resulted in a much-needed through goods road in the down direction that, together with service reductions due to line closures, by default alleviated the situation somewhat. Excepting one solitary visit to Shrewsbury shed, I failed to tread the streets of the cathedral city (a situation subsequently rectified) with its multitude of 400-year-old, black and white, half-timbered buildings, its winding steep narrow lanes, ancient passages, museums and castle. They were of little interest to me at the time. I was in the area for one thing: haulage on steam-operated passenger services.

Leaving Shrewsbury for the run northwards, trains are faced with a difficult start and, having passed Coton marshalling yard, begin the 1½ miles of 1 in 100. The climb continues for a further 2¾ miles past Leaton, the first of the derelict stations. Once the summit is reached, the line continues through undulating gradients until Rednal, where the 5-mile climb to Weston Rhyn begins. Shortly afterwards at Whittington, the Cambrian 'main line' (Welshpool to Whitchurch), over which I travelled in November 1964 but which closed just weeks later, crossed above.

After Gobowen, the climb to Weston Rhyn, formerly known as Preesgweene, resumes. Passing the site of the closed Trehowell Halt, the line enters a pleasant cutting, to emerge on Chirk viaduct, with Telford's aqueduct immediately to the east. A short tunnel precedes the entry to Chirk station, after which the line runs alongside the canal through the pleasant scenery of South Denbighshire and Cefn, and at the crossing of the Dee there is a magnificent view up the Vale of Llangollen framed by the Froncysyllte aqueduct. Indeed, Bradshaw eulogised at the time of his travels: 'To the tourist this line of railway holds out particular attractions. The Vale of Gresford, the grounds of Wynnstay, the valley of the Dee and the Vale of Llangollen offer the most beautiful view unsurpassed for grandeur and picturesque effect.'

After progressing through Ruabon, the former junction station for the 1965-closed Barmouth line, we now approach Wrexham, which was the centre of a complex network of railway lines and the busiest intermediate station, from which the ex GCR Wrexham Central station and Croes Newydd shed could be observed. Beyond Wheatsheaf Junction at Wrexham, the descent of the notorious Gresford 1 in 80 bank starts, with trains being required to observe strict speed limits due to mining subsidence. Once the 15mph (down) and 30mph (up) restrictions are passed, the trains can run freely over the level plain of the Dee estuary, with the approach to Saltney, passing the once extensive marshalling yard, usually being very fast. At the junction box, the former LMR boundary sign was noted before the approach to Chester, cutting through the rock of the city over quadruple track.

A day after John Lennon controversially claimed the Beatles were more popular than Jesus, I briefly put my feet on Shropshire soil, it being the second Saturday in March 1966. Merchant Navy 35027 *Port Line* had worked a 'footex' from Southampton to Wolverhampton but, rather than attend the football match, a far better way of occupying the hours prior to its return was undertaken by the half a dozen or so enthusiasts I was with.

Arriving into Wolverhampton thirty-four minutes late at 1420, we caught the following Birkenhead departure northwards. Having missed the preceding Paddington service, all we could manage was a tight connection at Gobowen before returning south for the home-going SR Pacific journey. On the northbound run, Brush Type 4 gave way at Shrewsbury to 4-6-0 45311, which, having inherited a fourteen-minute late-running train, luckily managed to maintain the schedule and arrived at Gobowen at 1607. We raced over the footbridge and jumped aboard the 1445 ex Birkenhead, with Llandudno Junction's 45277, which was brewing up, eager to depart.

What a run we were about to experience: eighteen minutes and two seconds for 18 miles, with Haughton being passed at 80mph and a maximum of 84mph between Baschurch and Leaton. This train went on to achieve celebrity status as Britain's last steam-hauled express timed at 60mph from start to stop; the mile-a-minute timing (no doubt scheduled by a pro-steam fan in the timings section at the LMR HQ) was booked over the 18-mile 4-chain 'racing' section between Gobowen and Shrewsbury. Not that we were aware of that at the time, with most of us struggling for breath after our exertions, and me red-lining two required Black 5s in my ever-present Ian Allan book.

With D1747 (again) taking over at Shrewsbury, we had a mere twenty-minute wait at Wolverhampton for the returning 'footex', and with word having spread among the enthusiast fraternity, the initial ranks of just six SR enthusiasts had swelled to twenty-six – our Birmingham bashing colleagues joining us for a ride with this SR interloper!

For those readers interested in football, although the match ended in a 1-1 draw, Southampton earned promotion from the Second Division at the season's end. Other football headlines that spring were Liverpool winning (for the second time in three seasons) the First Division, while in the FA Cup, Everton beat Sheffield Wednesday 3-2 in the final. Away from sport, Harold Wilson's Labour Government won an increased majority from one to ninety-six seats, Moors murderers Myra Hindley and Ian Brady were sentenced to life imprisonment, Longleat Safari Park was opened and, in the charts, the Walker Brothers' 'The Sun Ain't Gonna Shine Anymore', the Spencer Davis Group's 'Somebody Help Me' and Manfred Mann's 'Pretty Flamingo' all had a look-in at the top.

It wasn't until the third week in May that a return visit to the Shropshire steam scene was made. Rumours were circulating that the trains north of Shrewsbury were about to be turned over to DLs but no one knew when, the enthusiast grapevine not always proving reliable. Indeed, the line hadn't been a priority among us steam enthusiasts, with the 'guaranteed' longevity ensuring visits to other areas where steam was known to be finishing deemed more urgent. With the frequency of services, particularly northbound, not kicking in until after 1100, a regular 'killing time' move was to travel down on the 2345 Euston/Barrow 'kippers' to Preston and then the

0535 Crewe stopper from there to Wigan NW where, after an hour's fester, the Brit-booked 0610 Blackpool South to Euston was taken to Warrington Bank Quay before finally heading to Chester on the 0740 ex Manchester Exchange. So there we were, in situ, so to speak, to wait and see what was going on, with three catches already under our belt.

Below is a list of the trains covered during my visits to the line, together with the number of instances I was aboard for all or part of its journey. As can be seen, there was a dearth of northbound services during the morning but, by midday, one was able to station hop at will.

Birkenhead Woodside	0855	1145		1445	1630
Chester General	0933	1228	1435	1531	1720
Wrexham General	0956	1249	1456	1552	1741
Ruabon	1007	1259	1506	1602	1751
Gobowen	1017	1314	1520	1612	1805
Shrewsbury	1038	1333	1540	1632	1825
Paddington	1410	1705	1900	2015	2210
Instances	7	6	4	10	2
Paddington	0820	0910	1210	1310	1410
Shrewsbury	1134	1225	1532	1637	1723
Gobowen	1156	1247	1555	1659	1747
Ruabon	1209	1303	1607	1712	1803
Wrexham General	1216	1311	1615	1719	1810
Chester General	1238	1345	1648	1741	1850
Birkenhead Woodside		1417	1725		1925
Instances	3	9	8	2	4

My aim on each visit was to cover as many of the trains as possible shown in the above table and, having made my way to the western end of Chester station (where the Birkenhead's reversed) and dependent upon information received from the gaggle of 'platform-enders', who had either phoned the foreman at Shrewsbury and Chester sheds or observed earlier movements, I began to formulate my plans for the day. Luckily, on almost all occasions the 0933 departure was a required locomotive.

An often-repeated pattern of moves became the norm, sometimes travelling through to Shrewsbury or alternatively alighting at Wrexham and doubling back to Rock Ferry, courtesy of two DMU journeys, to intercept the 1145 ex Birkenhead Woodside. There was an infuriating imbalance of services off the Wirral, which sometimes resulted in an individual Birkenhead-allocated tank working just one passenger service all day.

Standing at the west end of Chester station and equipped with tickets to Wrexham and Rock Ferry, if, by lunchtime, nothing required was in circulation, refuge from the weather was sought in the buffet/refreshment room for a pie and a pint, with one of us intermittently dashing out for observational purposes. In those days, the bar always shut at 1500 and so, irrespective of the weather, the last few hours were spent, more in hope than anything else, out in all elements. Further variations, dependent on needs, sometimes stranded me at either Wrexham or Ruabon for an hour or so, having been caught by the three-hour gap in northbound steam services during the early afternoon. Although never experiencing it myself, with the locomotives being turned out at both Chester and Shrewsbury not always in the best condition, there were frequently reported substitutions at Wrexham with replacements being sent out from the nearby Croes Newydd shed and often resulting in an hour's delay!

Attempts to avoid the returning legs of locomotive duties also had its pitfalls, with the foreman at Shrewsbury shed, in particular, not keeping locomotives to their booked diagrams. With the day's travels almost over, a frequently used 'connection' was made at Shrewsbury between the 1445 ex Birkenhead (arriving at 1632) and the 1310 ex Paddington (departing at 1637). It was a sight to behold as, before the up train came to a stand, doors would open and a stream of enthusiasts would run through the subway and leap aboard the already signalled departure of the 1637. The first ones to the train held open the doors for the less fit of us and, with platform whistles and shouts of 'shut that door', only once did we not make it.

My first experience of this scenario was on Saturday, 21 May 1966 and, returning to London which was buzzing after Cassius Clay had beaten Henry Cooper at Arsenal's football ground, I was well satisfied with my catches of Birkenhead's Stanier Tank 42616 (which attained an impressive 72mph between Mollington and Upton-by-Chester), Speke Junction's

45059 and Chester's 45325, 45427 and 45438 together with its Standard 4MT 76095 – all duly red-lined in my *Loco Shed* book.

The Castles, Granges and Halls had long gone by the time of my visits, and although Stanier Jubilees, BR Britannias and 9F 2-10-0s had been reported of late, I consider myself fortunate in collecting runs behind six classes of steam locomotives – the obvious and most prolific of these being the Stanier 4-6-0 Black 5s. With the foreman at both Chester and Shrewsbury adept at using whatever was available, an over-expectancy of a continuous provision of visiting locomotives from other LMR depots was the 'norm'; this sometimes backfired having already caught them on their home patch.

The condition of the remaining locomotives left a lot to be desired and they were failed or withdrawn upon collecting even the smallest defect. Fortunately, some that were redundant from other depots in the north-west had been overhauled and put into storage – a strategic reserve perhaps? A number of these appeared on the line in the final few weeks, with the resultant combination of a relatively 'fit' locomotive and a keen crew sometimes leading to exhilarating runs.

The Riddles-designed BR 5MT 4-6-0s caught working the line were all (73040 of Croes Newydd excepted) Patricroft-allocated, the Chester foreman sending out both normal and Caprotti-fitted examples indiscriminately. Shrewsbury had an allocation of their smaller sisters, the BR 4MT 4-6-0s, mainly for use on the Cambrian Coast services and these were occasionally sent out on Chester trains. With their lower tractive effort, however, they often dropped minutes with slower acceleration away from stations on these tightly timed Class 1 trains.

The only chance of catching a run with any of Chester's 76xxx Standard 4MTs was the occasional use of them on the 15-mile leg to Birkenhead. With, by early 1966, their numbers down to eight, I was indeed fortunate in collecting runs with four of them, Croes Newydd-allocated 76048 being an added bonus.

By early 1966, Birkenhead shed had four Fairburn 2-6-4Ts – 42086, 42087, 42121 and 42156 – of which only 42086 survived into 1967, a transfer in from Bolton of 42133 in December 1966 ensuring she wasn't left alone. Turning to the Stanier variants, by early 1966, Birkenhead shed had six examples – 42548, 42587, 42606, 42613, 42616 and 42647. All of these saw in 1967, except 42647 as she was replaced by the transfer in of 42663 during January 1967 from Trafford Park. I caught her on the penultimate weekend of steam in the area in February 1967. Only 42587 and 42616 escaped the March 1967 cull by fleeing to Low Moor in Bradford.

Chester was without a doubt a pivotal point for a never-ending supply of steam locomotives from a varied selection of LMR depots. There were, however, occasions when the lure of the great many summer extras on the WCML meant I made just a flying visit to Chester, arriving in on the 0740 Manchester Exchange to Llandudno at 0903 and dashing over the footbridge onto the 0740 Llandudno to Manchester Exchange opposite-way working (0905). Literally just a minute or two on Chester's platforms! Saturday, 25 June was such an example: 73073 arrived at 0904, 73160 departed at 0906. Four weeks later, and I didn't even alight onto Chester's platforms, passing through westwards to Rhyl with Jubilee 45562

This and the following six photographs were all taken on Saturday, 21 May 1966. Here at Chester General, Patricroft's Standard 5MT 73011 is working that day's 0740 Manchester Exchange to Llandudno.

Here are two views of the 0740 'Club train' Llandudno to Manchester Exchange at Chester General. Perhaps emphasising their go-anywhere route availability, I had travelled with this Holyhead-allocated Stanier 4-6-0 44770 in the Bradford area just eight days earlier.

Despite meticulous planning and getting 'the word on the street' from platform-enders, there were still periods when photographing trains was more fruitful than finding a steam-hauled one to catch. Here BR 9F 2-10-0 92082 is seen reversing light engine on the north side of Chester station. Transferred from Saltley to Birkenhead in November 1963, she survived until 8H's closure in November 1967.

One of Chester's three surviving Jinties, 47389, was on station pilot duties that day. This 40-year-old Fowler-designed 0-6-0T 3F was withdrawn four days later.

Probably on a running-in turn out of Crewe Works, Warrington Dallam's 45375, in pristine condition, sits in a south bay platform at Shrewsbury.

In the bay platforms at Chester General, home-allocated 4-6-0 45427 has just brought me the 42 miles from Shrewsbury on the 0820 from Paddington. This Armstrong Whitworth-built 29-year-old was withdrawn just three months later.

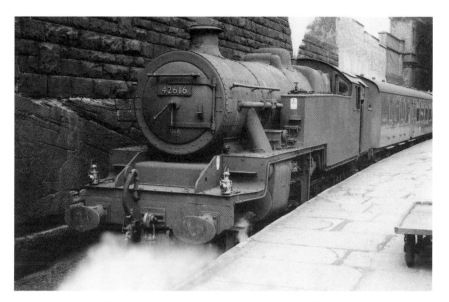

Twenty-nine-year-old North British-built Stanier 4MT 42616 (embellished with a chalked Carnforth shed code) waits time at Birkenhead Woodside with the 1445 for Paddington, which she will work the 15 miles to Chester. Transferred upon cessation of the Paddington services the following year to Low Moor, I was to catch a further run with her from Normanton to Halifax on the infamous 0210 York to Manchester Victoria.

Alberta on the Leeds–Llandudno summer Saturday train and returning to Manchester with Patricroft stalwart 73006.

As I am now moving into the high summer of 1966, perhaps a few items of non-railway news might be justified for inclusion here. That June, Barclays had launched the first British credit card, while the following month, with a record 32 million TV viewers, the country held its breath when England beat West Germany 4-2 to win the football World Cup. The highly disruptive eight-week strike by the National Union of Seamen, with family holidays ruined by the absence of cross-Channel ferries, finished in mid-July, while the controversial *Till Death Us Do Part* began its first series on TV. In the charts, the Beatles' 'Paperback Writer', the Kinks' 'Sunny Afternoon' and the Troggs' 'With a Girl Like You' were holding sway.

A section of the former GWML from Paddington to Chester that had by this time only one steam passenger service remaining was the 29-mile stretch from Wolverhampton to Shrewsbury. With the train involved, the 1510 Paddington to Shrewsbury, only running Mondays to Fridays, it meant a day's leave to cover the said track by steam had to be taken.

The Great Western Steam Retreat

The euphoria from England's World Cup win the previous Saturday was still being felt throughout the land when Friday, 5 August saw me departing Waterloo for Basingstoke to connect into the Poole to York train. This inter-regional service arrived at Basingstoke behind Black 5 44777, and having boarded the train I red-lined it accordingly, just prior to it being failed.* I learnt a valuable lesson that day. *Never* take for granted that you have caught a run with a locomotive until it moves! It was eventually replaced by BR Standard 5MT 73093 and a forty-one-minute late arrival at Banbury still allowed sufficient time to connect into the 1510 ex Paddington. This was taken forward from Wolverhampton by Oxley-allocated 44872 and, after an abortive two-hour wait at Shrewsbury in the hope of required steam, I departed for Crewe en route for a WCML bash the following day.

The Beatles were making headline news that month, having performed what was to turn out to be their last concert at Candlestick Park, San Francisco, while holding the top spot here with the double-sided 'Yellow Submarine'/'Eleanor Rigby'.

We now fast forward to the last week of the 1966 summer timetable. The outlook for steam-operated passenger services throughout the LMR was not good. With the short-dated usually steam-worked services ceasing from 5 September, I embarked on a five-night marathon bash throughout the north of England. On the first day, Wednesday, 31 August, I blitzed the North Eastern Region, subsequently making my way via Preston and Warrington to Chester on the Thursday morning. Unlike elsewhere, where it was a case of hoping the trains would be steam-hauled, at least this was going to be a stress-free day in respect of knowing that the majority of services would involve steam.

Starting with the customary 0933 departure and visiting Shrewsbury, Wrexham and Birkenhead, I departed from the area nine hours later, very happy after 167 miles behind six *required* locomotives, albeit coming across a *not-required* 45231 twice. The highlight of the day was Shrewsbury's BR 4MT 75004, which worked that day's 0820 ex Paddington and achieved a maximum of 71½mph between Shrewsbury and Gobowen. The locomotives caught on that visit were: Birkenhead's 42613 & 42647,

---

* Although missing out on a run with her at the time, I caught up with her in September 1967 when the by then Edge Hill-allocated locomotive worked the 0100 Manchester Exchange to Wigan NW.

It's Thursday, 1 September 1966 and, completely oblivious to her impending withdrawal just days away, the last surviving of seventy examples of Hawksworth-designed 0-6-0PT 2Fs, Croes Newydd-allocated 1628, potters around.

Stoke's 75030 scatters the pigeons when passing through Chester General with a westbound freight. She was withdrawn at Tebay in December 1967 upon the cessation of steam banking duties up Shap.

Shrewsbury's 75004, Chester's 45231, 76036 and 76044 and Oxley's 44805. Later, a visit to Crewe's all-night refreshment room was called for before a further two days were spent on the WCML.

Somewhat prophetically, with the Small Faces' 'All or Nothing' having its one week at the top on Saturday, 17 September, and following a DL-infested venture into Scotland, I arrived into Chester at 0950 on a DMU from Crewe, i.e. having missed the 0933 Shrewsbury departure. With no prospect of booked steam-hauled services for several hours, I was considering a Chester shed bunk when Patricroft's 73157 arrived in with the late-running 0830 (Relief) from Manchester Exchange to Llandudno, which should have departed at 0948. Problem solved, and having enjoyed a return trip to Llandudno Junction, upon arriving back into Chester at 1300 I restarted my normal activity of train monitoring in the hope of required catches. Even starting that late in the day, I was fortunate to collect six runs over four hours, visiting Ruabon and Rock Ferry before heading off to Wigan en route home to London. Locomotives caught that day were: Birkenhead's 42086, Chester's 44875 and 45031, Llandudno Junction's 45004, Edge Hill's 45307 and Croes Newydd's 76048.

Saturday, 22 October saw me embark on one further brief visit. Despite the emphasis on chasing Brits on their few remaining WCML services, I managed to call in on the Chester services for a five-hour sojourn. Travelling south on the 0933 departure, I left early enough to intercept the 1346 ex Barrow at Warrington. Even so, the five trains I caught reaped the following locomotives: Stourton's 45080, Crewe South's 44681, Chester's 76052, Stockport Edgeley's 44782, together with Patricroft's 73157 (again). The previous day had seen the tragic Aberfan colliery spoil tip collapse in which 144 people – 116 of them children – had died.

Two weeks later and, although the Beach Boys' 'Good Vibrations' was topping the charts, my own vibrations were not of the lucky kind when, working to the same train plan, out of the three locomotives caught – Birkenhead's 42587, Chester's 76052 and Croes Newydd's 45130 – only the last was required and, as a result of an annoyingly quick turnaround by the Shrewsbury foreman, I managed to catch her twice. At least on that fireworks night I was on board the last steam-operated 1930 Stockport to

21st October

Time	Place	No.	
2152	Smy	NR	✓
2227	E+C	—	✓
2350	Euston	E3089	

22nd October ——

↓	Crewe	D1631*	
0345	Preston	—	✓
0535	Preston	45250	IL
0607	Wigan NW	—	2L
0721	Wigan NW	44669	13L
0737	Warr BQ	—	14L
0829	Warr BQ	NR	4L
0903	Chester Gen	—	3L
0933	Chester Gen	45080	IL
1038	Shrews	—	✓
1134	Shrews	44681	✓
1155	Gobowen	—	✓
1247	Gobowen	73157*	6L
↓	Chester G.	76052	·
1417	Birk Wd.	—	2L
1445	Birk Wd.	44782	rL
1515	Chester G	—	?
1532	Chester Gen	NR	1L
1603	Warr BQ	—	2E
1608	Warr BQ	70005*	6L
1641	Crewe.	—	8L
1812	Crewe	E3138	14L
1851	S. Edge	—	4L

Time	Place	No.	
1930	Stockport E.	42124*	✓ 4l
21.16	Bradford E.	—	✓
2200	Bradford E	45197	IL 21
2253	Huddes.	—	5L
2325	Huddes	D281	22L 18
↓	Stalyb.	45411	· 8
0026	Man Ex	—	23L
0100	Man Ex	73127	IL
0136	Wig NW	—	6L
0253	Wig NW	D1838	27L 3
↓	Crewe	E3064	
0620	Euston	—	4E
06 09½	BF	NR	?·
06.44½	PW		·

	45
	76
	76
	45

Notebook extract detailing my Saturday, 22 October 1966 visit.

Having spent the first seven years of her short sixteen-year life on the SR, Fairburn 2-6-4T 42086's final shed was at Birkenhead. On Saturday, 5 November 1966 she is seen performing ECS duties at Birkenhead Woodside.

Bradford service for an exhilarating 47-mile ride over the Pennines with Low Moor's Fairburn 42116. Everywhere the growing number of redundant steam locomotives could be seen in sidings adjacent to depots, placed out of the way, disregarded and unwanted.

Retrospectively, at 35,575 miles, 1966 was to prove the zenith of my yearly steam mileages.

18

A Fabulous February Frenzy

Let's start this chapter with some non-railway news from those early weeks of 1967. Racing boat speedster Donald Campbell was killed on Coniston Water, the M1 was extended to Leeds, Milton Keynes was designated a 'new town' and the New Year's Honours went to England football manager Alf Ramsey and captain Bobby Moore for their contributions to the 1966 World Cup win. Music wise, just two songs held the top spot for those first weeks of 1967, the Monkees' 'I'm a Believer' being ousted after four weeks by Petula Clark's 'This is My Song'.

As for railway news, the Eastern and North Eastern Regions of BR were merged and the last steam locomotive to be out shopped, having received an overhaul in February at Crewe Works, was Britannia 70013 *Oliver Cromwell*.

The full Euston to Wolverhampton electric passenger services, having been inaugurated in April 1966, were to be implemented from Monday, 6 March 1967, and from then the London (Paddington) to Birkenhead (Woodside) services would run no more. With sufficient suitable DLs 'found' to operate the Cambrian Coast services, the ready supply of steam-hauled passenger services centred on Shrewsbury and Chester would no longer be available to steam-haulage bashers such as myself. Word of the exhilarating performances on the Shrewsbury to Chester line, with high-speed running, had spread far and wide and the line had become a magnet for enthusiasts.

Frequently crossing paths at various locations throughout the railway network, the most regular of us were 'awarded' nicknames that predominantly reflected our residence – i.e. Prestatyn, Midland Red, Andover and Bedford – although many others (unable to be printed here) were also used.

Although it was a sad day when the line north from Banbury was wrestled away from the WR into the hands of the LMR in January 1963, to the steam enthusiast, as the LMR was less zealous in ridding itself of steam, it was a saving grace, at least for a while longer. Apart from the sporadic usage of DLs, the LMR left this section alone, superficially therefore (certainly infrastructure wise) retaining much of its GWR heritage. Coupled with the fact that the line had been passed for 90mph running, together with the incongruous decision to divert any remaining freight traffic away, there were fewer chances of delays to passenger traffic. The grapevine was abuzz with rumours of a last fling and sure enough a 93mph speed was actually recorded in January 1967 aboard the 1445 ex Birkenhead with Driver Bernard of Shrewsbury, who was fast becoming as well known for his exploits as Nine Elms Driver Porter. Indeed, the Chester and Shrewsbury footplate crews certainly entered into the spirit of the final weeks with some outstanding performances with locomotives not always in the best condition. The financial incentives in the form of whip-rounds offered up to astounded drivers, requesting them to 'have a go', certainly assisted matters.

Once it became known that the line speeds were being shattered, the floodgates opened for more, and although the 'bad' runs outnumbered the good, anticipation is the steam enthusiast's adrenalin and you never knew when the next good run was going to occur. During that final year, both Chester and Shrewsbury motive power depots were dirty, unkempt, unloved places of work with equally run-down locomotives.

In many ways it was unfair to expect men who were on diesels for much of the time to be subjected to the pitch and toss of a Class 5 trying to maintain schedules calling for high speeds. It says much for the quality of the crews and shed staff at both depots that such brilliant, fast-running work was obtained from machines with little more than a year to live. The steam crews knew their jobs had become short-lived, as did those throughout the UK. No doubt some enjoyed participating in making railway history and were perhaps aware that, with so many people recording their exploits, their names would be documented by the railway press.

A Fabulous February Frenzy

My first visit to the line that February saw me make my way to the predictable 0933 departure via the 0100 Manchester/Wigan sleepers and the 0535 Preston to Crewe. The date was Saturday, the 11th and I took the 44715 (Crewe South)-worked train to Wrexham, doubling back on two DMU services to Rock Ferry for the 1145 ex Birkenhead. Usually worked by a Birkenhead tank, 44814 was not exactly welcomed by me as I'd travelled with her on the Marylebone services two years previously. Matters were to improve, however, when upon reversal at Chester, Shrewsbury's Black 5 45132 took over. Alighting at Wrexham after a mere twenty-minute wait, the 0910 ex Paddington had 6A's 44917 in charge. Upon arriving into Chester with just enough time for a pie and a pint, I then boarded the 1435 starting service from there to Wrexham, worked by home-based 45298. There was now an hour to kill and so we (there was a group of about half a dozen of us) decided to bunk Croes Newydd shed, following the instructions provided by Fuller's *British Locomotive Shed Directory*:

> The shed is in the triangle formed by the Ruabon – Wrexham – Brymbo lines. Leave Wrexham General station by the main ext. Turn left into Regent Street and right into Bradley Road. Turn right into Watery Road and a path leads to the shed over a boarded crossing from a gate on the left-hand side, just past the level crossing. Walking time fifteen minutes.

11/2/67	Croes Newydd @ 1515
In steam	45130, 48122, 48252, 48253, 48440, 48665, 48669, 75002, 75052, 76040
Dead	44776, 44872, 45198, 48325, 48632, 48697, 75021 (6D), 75046, 75071
Withdrawn	4646 (2A), 9641, 48385 76048
Diesels	D325, D339, D342

With no one stopping us, we wandered around, carefully picking our way among the detritus of a working shed not helped by the sulphurous gloom occasionally illuminated by shafts of light penetrating the soot-blacked skylights. One abiding memory is of the heightened sense of size when viewing the 5MT and 8Fs from ground level rather than from the platform level they were more usually seen from. A Bryan Hicks photograph in *Steam*

World magazine of a filthy 44832 on the turntable at Shrewsbury emphasises it was at the 'fag end' of LMR steam: 'With water filled (blocked drains) pits together with the crew having to manually crank handle the turntable it was hard to imagine that locomotives kept in conditions such as this were storming along the Gobowen to Shrewsbury racetrack at 80+ mph.'

Returning in time to catch the 1552 for the hop down the line to Ruabon, with a mere five-minute connection, the southbound train would only have been travelled on if a 'required' locomotive was working it. Having been pleasantly surprised by Patricroft's Caprotti 73135 turning up, it was a must – the 'connection' at Ruabon was made with seconds to spare!

With yet another required 4-6-0, 44713, in charge of our train into Chester at 1632 and knowing that Stanier 42647 was booked to take the train forward, I thought that was the end of the day's catches. However, the tank had been failed and so a trip to Hooton was made where 45285, having been taken off a parcels service, backed on. Still managing to make what was to become over the next few weeks the regular homeward-bound train out of Crewe, the 1850 for Euston, it was with great satisfaction that I was able to red-line eight entries in my *Loco Shed* book – the best for many months. Indeed, the underlinings within the said publication were looking less sparse now. No longer were there isolated red lines here and there but unbroken chunks of them, documented evidence of a seasoned haulage basher. So, the pattern was set for the following three weekends: hopping on and off services at intermediate stations dependent upon when required locomotives materialised.

The following Saturday (18th) saw us arrive in the area courtesy of Brit 34 *Thomas Hardy* on that morning's 0610 ex Blackpool South (taken to Warrington). Although Patricroft's Standard 5MT 73159 was a bonus on the 0933 departure, it was a barren three hours before any further required locos were caught. On top of that, the infamous five-minute connection (1632 to 1637) at Shrewsbury was not made because the now Stoke-allocated Black 5 44770 was not in the best of health with the inward working, and so I had an hour's fester for Brush Type-4 D1913 to Crewe en route to the 1850 London-bound train. The other locos caught that day were Speke Junction's 44725, Chester's 45369, Edge Hill's 45005 and Stoke's 44713.

The Chester foreman had borrowed visiting Patricroft Caprotti 4-6-0 73159 on Saturday, 18 February 1967. She is seen waiting in the centre road ready to work the 0855 ex Birkenhead Woodside, upon reversal, forward to Shrewsbury. This frequent habit of the foreman was a delight to a haulage basher, with 'foreign' locomotives regularly being put out on these trains.

Later that day, at Wrexham General, Croes Newydd-allocated BR 4MT 75002 storms through with a southbound freight. She was transferred to Stoke upon 6C's closure to steam that June.

Another scene at Wrexham General sees Horwich-built Stanier 5MT 44713 departing with the 1145 Birkenhead Woodside to Paddington. Transferred to Lostock Hall that May, she survived until the very end in August 1968.

17th February 1967	via st mary cray/victoria			
0816	Clapham Junction	✓	82019	L84
0824	Kensington (Olympia)	1E	—	34052 ~1700
0833	Kensington (Olympia)	2E	82019	Freehter 17/2
0841	Clapham Junction	3E	—	
	via Wimbledon West			
1723	Waterloo	2L	34004	
1907	Southampton Central	✓	—	
1914	Southampton Central	7L	35012	
2051	Waterloo	3L	—	
2220	Euston	✓	E3102	70053 1825
18th February 1967				70023 1346 Bm
0122	Crewe	✓	·	70052 Southbound pands
0225	Crewe	3L	D1353	
0345	Preston	2E	—	46120 then cleaned up for LCGB railtour
0555	Preston	✓	NR	for Workington
0607	Kirkham + Wesham	1L	—	
0640	Kirkham + Wesham	✓	70034	
0737	Warrington BQ	4L	—	P
0829	Warrington BQ	1L	NR	73159 Chester
0903	Chester General	✓	—	
0933	Chester General	✓	73159	
0956	Wrexham General	✓	—	
1039	Wrexham General	✓	NR	
1100	Chester General	4E	—	
1108	Chester General	✓	NR	
1132	Rock Ferry	3E	—	

18/2/67 (cntd)				
1152	Rock Ferry	✓	45005	
	Chester General	↓	44713	
1249	Wrexham General	2E	—	
1311	Wrexham General	8L	44821	
	Chester General	↓	45369	
1359	Hooton	3L	—	
1405	Hooton		NR	
1422	Chester General	✓	—	
1435	Chester General	1E	44725	
1456	Wrexham General	✓	—	
1552	Wrexham General	✓	44770	
	Shrewsbury			
1632	Shrewsbury	5L	—	
1723	Shrewsbury	9½	D1943	
	Crewe			
1801		22L	—	
1850	Crewe	1L	E3157	
2100	Euston.	3L	—	
	via victoria /st mary cray			

A Fabulous February Frenzy

For the third Saturday in a row (the 25th), having travelled via Preston and catching 5B-allocated 45021 on the all stations 0535 through to Crewe, I had sufficient time to get to Rock Ferry for the 0855 ex Birkenhead – i.e. the 0933 forward from Chester – rather than see it arrive into Chester with a required loco that often never reappeared again that day. Thinking, for whatever reason, all 8H's tanks had been withdrawn, I was overjoyed to have Stanier tank 42663, a recent transfer from Trafford Park, on the train. She was clearly in fine fettle as a maximum of 64mph was achieved south of Mollington. Three return trips from Chester to Wrexham during the day yielded several required Black 5s and the lunchtime sightings of two unwanted BR Caprottis were compensated for by Croes Newydd's BR 4MT 75021 on the 0910 ex Paddington. All in all, not a bad day's work. Other locos caught were Chester's 44771, Patricroft's 73127, 73139 and 73158, together with Shrewsbury's 44800. However, I wasn't finished with the Shrewsbury steam scene that weekend because after just a few hours at home I headed back on the Severn & Dee Rail Tour.

19

The Severn & Dee
Rail Tour

After about six hours in a warm, comfortable bed, I set forth once again for the Shrewsbury area on Sunday, 26 February, the sole incentive being a rail tour that promised a run with a Crosti-boilered BR 2-10-0. Although I'd had a ride with a conventional 9F out of Marylebone in August 1966, the attraction of the use of one of Birkenhead's sub-class was too good to miss.

The year 1955 saw ten 9Fs, 92020–29, constructed with Franco–Crosti boilers – which incorporated a combustion gas-feed water heater for recuperating residual heat. In the event, the benefits of the experiment were disappointing, and so, with the Modernisation Plan leaving no doubt that steam traction had no future, the pre-heaters on these engines were sealed off in 1959 and later removed for conventional operation. With the pre-heater on these locos gone, the boiler barrel was therefore of a smaller diameter than that of the regular 9F fleet and the ten locos were thus reclassified as 8F.

The E3035-hauled nine-vehicle tour departed out of Euston promptly at 0930 and, after arriving at Nuneaton Trent Valley, handed the train over to a remarkably clean Oxley-allocated Stanier 4-6-0 44944 (required to boot!). We then wended our way through myriad lines, parts of which were freight only, via Water Orton, Sutton Park and Walsall to Wolverhampton High Level.

Oxley's Blackie 44944 comes off the Severn & Dee Rail Tour at Wolverhampton High Level on Sunday, 26 February 1967, on which she had worked the 33 miles from Nuneaton.

Having had a serious clean since the November 1966 tour, 7029 *Clun Castle* is at Wolverhampton High Level about to work the 72-mile leg to Chester.

I was only aboard this Severn & Dee Rail Tour for its advertised usage of a Franco–Crosti 8F. At Chester General, Birkenhead's 92026, spruced up for her fifteen (actually forty-one!) minutes of fame, is seen about to work the 21 miles to Crewe.

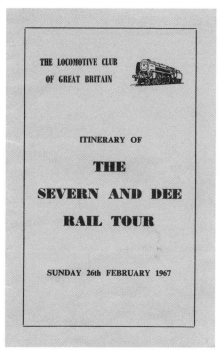

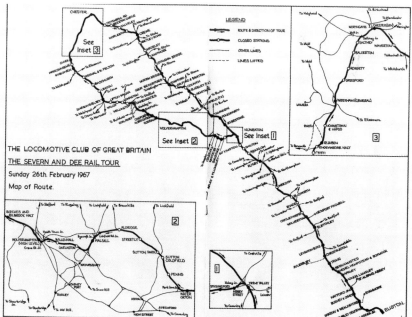

The Great Western Steam Retreat

There the preserved Castle 4-6-0 7029 *Clun Castle*, now a lot cleaner than my previous encounter with her (see p. 158), took us forward the 72 miles down the former GWR route, maxing at 78mph between Allscott and Upton Magna, to Chester. We passed through Shrewsbury without stopping, a rarity in itself, before making a photographic call at Ruabon and a water stop at Wrexham General.

Back then, you never knew if locomotives booked for these tours were going to appear. All it needed was a minor fault and they would be replaced at short notice by a.n.other! Crossed fingers became uncrossed upon setting my eyes on a resplendent 92026. I couldn't stop taking photographs of her. It didn't matter that it was only 21¼ miles from there to Crewe. It didn't matter that she lost seven minutes, maxing at a mere 47½mph on an easy thirty-one-minute schedule, later stated as being due to brakes dragging on the buffet car. What mattered was a – relatively expensive, mind – red line in my book, she being my 600th BR steam locomotive (for haulage).

Taken forward from Crewe by E3130, ten minutes were gained on the schedule and we arrived into Euston at the respectable hour of 1910.

20

The Great
Western Survivors

Here, in this colour section, I present a selection of the fifty-five (so far) Swindon-built machines I have caught in preservation.

I feel cheated at having been born just a few years too late to witness steam in the 1950s and am envious of those who did. Nevertheless, I am grateful for those I did manage to travel with and document. I certainly miss the comradeship of like-minded friends and colleagues from those wonderful days. For sure, resulting from my publishing exploits, many have surfaced at galas and via Facebook but with the purpose of our hobby, that of hunting down the fast-disappearing Iron Horse, having been taken from us, the common thread binding us together has gone. We grew up and matured together, subsequently going on to lives so different. Whenever those with whom I remain in contact meet up at various social gatherings (or regrettably at wakes), as soon as a particular event is mentioned out come the memories. Being stranded overnight at a station, just missing out on a good run, the euphoria at catching a run with the final remaining required survivor of a class – all good fun, and with the banter being conducted in a light-hearted manner, those days are fondly remembered.

My parents were of a generation who created, secured even, a lifetime of peace for us – allowing us to further our lives in whatever way we wanted. With greater affluence, revolution in communication, technical advances and pensions, to quote Harold Macmillan's statement, 'We've never had it so good'.

What a lot of fuss over locomotive liveries these days. Personally, I prefer to see them as I remember them in everyday life: work stained, unkempt, but still able to perform whatever task was asked of them. There was no nostalgia then: they were dirty and uncared for, unwanted even. The present-day owners are entitled to do what they want with them – they have invested their time and money. I am forever grateful that so many have survived the years for future generations to enjoy. The exploits of both *Tornado* and *Flying Scotsman* have enhanced the public interest in the preservation movement.

More than 130 former Great Western or GWR-designed locos have been saved from the cutter's torch, thanks predominantly to Dai Woodham, thus allowing me to continue my hobby of *still* catching required haulages.

My wife sometimes opines that my interest is an obsession. I closet myself away in an 'office' during the winter months (having satisfied the twice-daily needs of my dog) and tap feverishly away on articles (many of which never see the light of day) or on a book destined to be torn apart during the editing process at the publisher. I am gradually emptying my memory box of my travels; hopefully they will be of interest to all.

The Great Western Survivors

The last steam locomotive constructed in Britain, BR 9F 2-10-0 92220 *Evening Star*, is on the country end of the Old Oak Common shuttle on Sunday, 20 September 1981. Built at Swindon in March 1960, she was withdrawn at Cardiff East Dock after just four years and ten months and is currently a static exhibit at the National Railway Museum in York.

At the London end of the shuttle, run in connection with an open day at 81A, was Castle 4-6-0 5051 *Earl Bathurst*. She had been withdrawn at Llanelly in May 1963 and now resides at the Didcot Railway Centre.

The sole survivor of the once seventy-strong class, 0-6-0PT 1638 is seen at the South Devon Railways Totnes Riverside station on Sunday, 16 July 1989. She now resides at the Kent & East Sussex Railway (K&ESR).

Sunday, 22 October 2006 and at Sheffield Park a pair of veterans are seen having worked a train from Horsted Keynes. Nearest the camera is the Bluebell Railway's Dukedog 9017 *Earl of Berkley*. This 2P 4-4-0, the only survivor of a twenty-six-strong class, had been withdrawn at Oswestry in 1960. Furthest from the camera is visiting 3440 *City of Truro*, currently a static exhibit at the National Railway Museum in York.

Another Bluebell railway scene, this one at Horsted Keynes, sees visiting Llangollen-based 2-6-2T 4MT 5199, withdrawn at Gloucester in March 1963, arriving on Saturday, 21 June 2008 with a northbound train.

Sunday, 6 July 2008 and BR-built 9466, withdrawn at Radyr in July 1964, arrives into Wymondham station on the Mid Norfolk Railway with a train from Dereham.

The North Yorkshire Moors Railway-based, Collett-designed 0-6-2T 6619 visited the K&ESR on Saturday, 2 August 2008 and is seen here at Bodiam.

Suitably embellished with 'end of steam' graffiti to commemorate the cessation on BR forty years earlier, 80-year-old 2-6-2T 5542 readies herself for departure at Cheltenham Racecourse station on Saturday, 10 August 2008.

Beautifully restored former Cardiff East Dock-allocated Modified Hall 7903 *Foremark Hall* (an eighteenth-century Derbyshire house now in use as a preparatory boarding school) arrives into Winchcombe on Saturday, 10 August 2008 with a Toddington-bound service.

Privately owned Collett 4-6-0 5029 *Nunney Castle* arrives into Ropley with an Alton train, on Saturday, 13 September 2008.

On a typically wet Wednesday in August 2009, and at the Dean Forest Railways Parkend station, Pannier 9681, withdrawn at Cardiff East Dock in August 1965, runs around her train.

Hawksworth 0-6-0PT 1501 guests at the North Norfolk Railway's 2015 Spring Gala, seen on Friday, 6 March at Sheringham. After twelve years of usage by BR she was withdrawn at Southall in 1961 and then sold to the National Coal Board for a further eight years of service.

Collett-designed 1P 0-4-2T 1450, withdrawn at Exmouth Junction in May 1965, stands at Norton Fitzwarren on Wednesday, 21 March 2012, having worked the 1410 from Watchet, which then returned as an autotrain to Bishops Lydeard.

Withdrawn at Oxford in June 1964, 1944-built Hawksworth 6960 *Raveningham Hall* is seen at Minehead on Thursday, 22 March 2012 with the 1400 departure for Bishops Lydeard.

Seventy-year-old Class 2884 2-8-0 3850 is captured in camera on Thursday, 22 March 2012 when calling at Blue Anchor while working the 1000 Minehead to Bishops Lydeard. This 8F had been withdrawn at Croes Newydd in August 1965.

One of nine preserved Manors, 7820 *Dinmore Manor* departs Washford on Thursday, 2 October 2014 with the 0950 Bishops Lydeard to Minehead. This privately owned locomotive was one of the many withdrawn at Shrewsbury in the November 1965 cull.

The author with his 'window hanging' equipment at the 2014 WSR Autumn Gala.

They say patience is a virtue. After a fifty-two-year wait, Churchward's Manor 7822 *Foxcote Manor* was ensnared when visiting the K&ESR on Saturday, 6 May 2017. This was the locomotive viewed at Sutton Bridge Junction working the down Cambrian Coast Express having bunked 6D in October 1965.

Saturday, 14 July 2018 sees visiting 6023 *King Edward II* at Kingswear having worked in on the 1000 from Paignton. This 88-year-old 4-6-0 was withdrawn at Cardiff in June 1962 and is normally to be found at Didcot Railway Centre.

The same day and L94 (7752), visiting the Dartmouth Steam Railway from Tyseley, prepares to work the 1125 away from Kingswear to Paignton. Coincidentally also 88 years old, she was purchased by London Transport in 1959 and was taken into preservation after twelve years' service on ballast trains around the capital.

The author stands next to 100-year-old 2-8-0 2857 at Bewdley (both still going strong!) on Thursday, 20 September 2018 while she prepares to work the 0830 departure for Bridgnorth during the annual Severn Valley Gala.

Withdrawn by BR in 1959, Pannier 7714 spent fifteen years working for the National Coal Board before finding her forever home at the Severn Valley Railway. She is seen here at Kidderminster on Thursday, 20 September 2018, having arrived with the 0930 from Bridgnorth.

Swindon-built 59-year-old 9F 92214 spent the majority of her short five years and ten months in service at Welsh sheds. Now named *City of Leicester*, she is seen at Rothley participating in the 2018 GCR Autumn Gala.

It's Thursday, 11 April 2019 and visiting L92 rests at Chinnor having worked in on the 1410 from Princes Risborough. This 89-year-old Pannier (BR number 5786) had spent her working life in South Wales, being withdrawn in 1958 at Cardiff and sold to London Transport for use on ballast and engineering trains. Sold into preservation in 1969, she finally found her forever home on the South Devon Railway in 1993.

After catching a run with L92, I alighted at High Wycombe and noted the Grade II listed, 1854-built Brunel train shed terminus there. In use for a mere ten years, it became a goods shed upon the opening of an adjacent station on the new through line to Princes Risborough and beyond.

21

And Finally:
The Last Weekend

The final days of steam services on the Shrewsbury, Chester to Birkenhead line prob-
ably attracted more railway enthusiasts than any other event in railway history. The
day 4 March 1967 will almost certainly go down as one of those special ones. The
large crowds at Birmingham Snow Hill and Chester General had to be seen to be
believed. Public interest on this occasion was on an incredible scale. *

For the final weekend of Shropshire steam I took two days' leave, the pivotal point of the trip being centred on the last steam-hauled up Cambrian Coast Express from Aberystwyth to Shrewsbury on Saturday, 4 March. With the Euston to Birmingham line in chaos following a train crash at Stetchford, I travelled north, on Thursday, the 2nd, on the 0910 departure out of Paddington. It was normally booked for steam haulage from Shrewsbury, but Brush Type-4 D1722, an unexpected diesel substitution, took it forward. Travelling only to Gobowen, I returned south with a required Shrewsbury 45310, a locomotive that upon its transfer to Newton Heath was to monopolise the much-frequented overnight Calder Valley service the following year, which attained 79mph south of Leaton. Unlike many others, I had not been timing my services caught, which, taking into consideration some excellent fast running, was perhaps a mistake.

★ Chris Magner, *Steam's Finest Hours*, 2018.

Thursday, 2 March 1967 and Saltley's 8F 48220 makes her way through Shrewsbury station en route to the shed.

Upon arriving back into Shrewsbury, I learnt of the day's motive power on the Cambrian Coast Expresses, namely 75048 on the up and 75021 on the down. Would the required 75048 work the 0410 departure tomorrow? I was planning to be aboard it. Returning back down the line to Chester yielded two further catches of Chester's 45042 and Rose Grove's 44690 before heading to Crewe for the Brit-worked 1905 ex Euston, which throughout the last week of regular WCML steam was monopolised by 70016 *Ariel*.

Having sated my appetite at Crewe's all-night buffet, I boarded the 0225 train to Aberystwyth during the early hours of Friday, 3 March. During the forty-nine-minute layover at Shrewsbury, the Type-2 D5145 gave way to 6D's BR Standard 4MT 75033. With just a handful of passengers on board, I had a compartment to myself throughout the three-hour journey. At the calling points en route papers and mail were unloaded, with a lot of friendly banter being exchanged between train crew and station staff. Occasionally glancing out into the chilly darkness and catching glimpses of the locomotive's fire dancing off the hillsides, I closed my eyes and was lulled to sleep listening to 75033's exhaust as we headed west.

Arriving four minutes late into Aberystwyth at 0715, and with time to kill before the return 0955 departure (the Pwllheli portion previously travelled from Morfa Mawddach eighteen months earlier now being a connecting DMU), perhaps it's an opportune moment for a spot of history. The original 1863-built station was greatly extended by GWR in 1925, being provided with a grand terminus building and five platforms. With the decline of railway usage (in particular the Carmarthen services, which had ceased two years earlier) and of local tourism, the facilities were far too large for their purpose. The railway yard was lifted in the 1980s and the row of shops in front, known as Western Parade, was demolished in the 1990s to allow construction of a new retail park and bus station. The 1925 station building has seen several uses, including as a local museum, but was eventually sold off and converted into the Yr Hen Orsaf (The Old Station) Wetherspoons pub (it would have been nice to have had a high-calorie breakfast back then on that somewhat breezy cold morning!). This conversion maintained the architecture and has won awards. Other parts of the building have become an Indian restaurant, office space and accommodation for a local furniture recycling scheme.

Departing promptly at 0955, arrival into Shrewsbury was nine minutes early (1235), allowing me to connect into the late-running 0910 ex Paddington (1240), which, unlike the day before, was steam hauled, albeit with an already travelled with 45042. A few trains later, having unluckily encountered Black 5 45310 twice more thanks to a quick turnaround by the Shrewsbury foreman, I relished the catch of the day, the recently transferred in (from Bolton) Fairburn 4MT 42133 on the 1850 Wirral-bound departure out of Chester. Once again it was the same schedule as twenty-four hours earlier in that, following another run with Brit 16, 'supper' was taken at Crewe; what would we enthusiasts have done without such all-night provision of food?

So now it's Saturday, 4 March and a couple of non-railway facts can be thrown in. It was on this day that the first North Sea gas was pumped ashore and Queens Park Rangers became the first Third Division club to win the League Cup at Wembley, beating West Bromwich Albion 3-2. In the charts, Engelbert Humperdinck began a six-week tenure at the top with 'Release Me'.

Saturday morning and there were far more passengers aboard 1M41 to Aberystwyth. Type-2 D5075 spluttered its way through the moonlit countryside over the 32 miles to Shrewsbury, where 4MT 75033 was waiting *again*. I suppose I shouldn't grumble but there were several Shrewsbury-based 75xxxs I still required. The previous morning's scenario was recreated,

With makeshift numberplate and typically dreary external cleanliness associated with the final days of steam, Shrewsbury's BR 4MT 75033 waits at Newtown while working the penultimate steam-hauled up Cambrian Coast Express on Friday, 3 March 1967.

Vulcan Foundry-built, 32-year-old, Chester-allocated 45042 is about to depart from Gobowen later that same day with the 0910 Paddington to Birkenhead Woodside.

And Finally: The Last Weekend

The penultimate steam-worked down Cambrian Coast Express (she also worked the last the following day) departs Shrewsbury with Croes Newydd's 75021. Finding her way to Carnforth, she was withdrawn at just 14 years of age in February 1968.

perhaps this time with a greater poignancy as it was, after all, the last steam-hauled one.

Things were a little different that day with members of the Master Neverers Association (MNA) setting about the 4-6-0 during the turna-round at Aberystwyth, transforming the world-weary Standard 4MT into a respectable condition. Having manufactured a very lifelike replica number-plate for 75033, together with undertaking several hours of elbow grease and applying white paint to the smokebox stays and buffers, when adorned with the headboard, the locomotive looked stunning. Photographs of it in this condition have subsequently appeared in many publications over the years.

Beyond Machynlleth, the valley of the River Dovey initially offers the footplate crew a good chance to get established, the first mile or so being downhill or flat, but then the climb begins, and it gets progressively harder in stages over the next 13 miles. There were indeed many photographers lining the route that day, particularly witnessing the stiff climb to Talerddig, and as I was leaning out of the last window of the leading vehicle I was in many of their photos! The sense of the occasion coupled with the enthusiasm of the crew made it a memorable journey, with speeds of 55mph approaching

Twenty-four hours later and, courtesy of the MNA organisation, what a transformation. It's Saturday, 4 March 1967 and the 0955 Aberystwyth to Paddington, the final westbound steam-hauled Cambrian Coast Express, sees 75033 readying herself for departure. Shrewsbury was her eighteenth shed in her twelve-year life. She was transferred to Croes Newydd the following week and ended her days at Carnforth at the year's end.

Borth, 16mph climbing Talerddig and 63mph approaching Montgomery and again at MP 8. It was without doubt a fitting end to regular steam in Wales.

After arrival into Shrewsbury there was a 'what shall we do now' atmosphere and a great majority opted to return on the down Cambrian Coast Express. Having ascertained it was 75021 *again*, I wasn't bothered about that move. Instead, I was still hanging my hat on possible catches on the Chester trains.

Although aware that two rail tours were running that day, I had no knowledge of precise timings and when, at 1300, the recently restored 4079 *Pendennis Castle* stopped for a crew change on the Birkenhead Flyer, I jumped aboard. I was quickly searched out by observant officialdom and was quite happy to stump up the £10 single I negotiated to Chester. With an eleven-coach load, the Castle gradually lost time en route and only managed a maximum of 70mph north of Rossett.

The two rail tours were both run by the Ian Allan publishing organisation and, with 7029 *Clun Castle* on the preceding Zulu tour, there were fields full of photographers and families waving and applauding as we passed by. The tours were taken forward, albeit as a reduced seven-vehicle formation because of platform lengths at Birkenhead, by Standard 5MTs.

And Finally: The Last Weekend

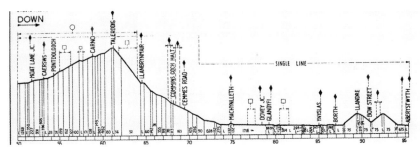

Gradient profile of the Cambrian main line.

With the opening of the L.M. electrification to Birmingham in March 1967 the

LAST THROUGH EXPRESS SERVICE

will operate between

PADDINGTON and BIRKENHEAD

To mark the occasion a

COMMEMORATIVE SPECIAL TRAIN

will be run on

SATURDAY 4th MARCH 1967

Departure from PADDINGTON approx. 8.55 calling at BANBURY, BIRMINGHAM (SNOW HILL), WOLVERHAMPTON, SHREWSBURY, CHESTER. **STEAM HAULED** between BANBURY and BIRKENHEAD—"CASTLE CLASS" BANBURY-CHESTER. This may be a last opportunity to visit

BIRMINGHAM (Snow Hill)

which may be closed shortly afterwards

Passengers may join the train at

	LONDON	BANBURY	BIRMINGHAM
2nd Class fares.	75/–	70/–	60/–

1st Class accomodation available at 99/– from any point.

BOOK EARLY "BIRKENHEAD SPECIAL" TERMINAL HOUSE SHEPPERTON MIDDX.

Advert in December 1966 issue of *Railway World*.

```
4079 / 11c / - (373)
Shrewsbury    0.00.
Coton Hill NR. 2 20.    28
keaton        8 30    38/58
Basehunk.     13 14.   51/64
Haughton.     17 32    60
Rednal        19 03    60/64
Whittington   22 12    54/47
Gobowen.      24 25    51½
Western Rhyn. 26 36.   60½
Chirk.        27 48    53
Whitehurst Halt 29 08   59
Cefn          30 34    48/43
Ruabon        33 03    46½
J + H         34 55    56/20
Wrexham Gen.  39 56.   22.
United Colliery. 42 40  35
Gresford Halt  44 33    39
Rossett.       46 27.   62/70.
Balabafon      49 23    69
Saltney Dee.   51 48.   24.
Saltney Jn.    53 20.   38.
               55 37
               52 45
Chester Gen    62 32
```

Both had started out of Paddington behind diesel-hydraulics but they took different routes as far as Banbury: the Zulu went via High Wycombe and gained its Castle at Banbury, while the Birkenhead Flyer departed from Paddington fifteen minutes earlier and travelled via Didcot where *Pendennis Castle* was waiting to take over. The two routes merged at Aynho Junction. North of there the Zulu was the leading tour, thanks to its more direct route, as used by the latter-day Paddington to Birkenhead expresses.

The renowned railway historian C.J. Allen was quoted as having said that he thought it little short of amazing that nearly 1,000 railway enthusiasts were prepared to pay a substantial fare for the privilege of travelling a long distance once again behind steam power. Perhaps the added attraction of the *Pendennis Castle* working forward from Didcot, being heralded as the very last steam train running over WR metals, might also have contributed to the tour's popularity. I was unaware of it at the time but a BBC *Tonight* current affairs crew were aboard and broadcast the results the following Monday.

Arriving into Chester, there were amazing scenes as throngs of people packed the platforms and, much to the British Transport Police's consternation, all over the running lines and sidings. Indeed, I read afterwards, there was also mass trespass at Banbury and Birmingham Snow Hill* on adjacent running lines, ignoring the fact that service trains were operating over them. Scenes reminiscent of rugby scrums using elbows and feet in order to obtain the best shots were the norm – and present-day readers think, in connection with the main-line *Flying Scotsman* outings, that trespass is something new!

I left the Birkenhead Flyer at Chester, having noted what would turn out to be my only other catch of the day, BR Standard 5MT 73097 on the 1435 Chester to Paddington service. Alighting from this at Wrexham to catch the following 1445 service from Birkenhead with 44917, we were treated, courtesy of Driver Bernard of Chester, to a splendid seventeen-minute, eighteen-second, start-to-stop Gobowen to Shrewsbury run. This included six minutes of 80mph-plus, maxing at 85mph south of Leaton, the locomotive having been specially diagrammed, as being the best mechanically, courtesy of Chester staff. Someone did a passenger count of the train and out of the eighty-two persons aboard, forty-five were enthusiasts!

Then, for the very last time, together with a stream of other enthusiasts, we dashed through the subway, just making the tight connection into the 1637 departure back north to Chester. Taking over from the incoming diesel, green-liveried 73097 had been turned round in a very short time to work the Shrewsbury to Chester leg of this Paddington to Birkenhead service. Initially disappointed by that fact, I was soon made to appreciate what a wonderful steam journey I was about to enjoy.

The corridors were packed and, after a violent slip, quick acceleration was achieved with Haughton being passed at 82mph and Gobowen reached in eighteen minutes and forty-three seconds. With Driver Webb of Shrewsbury in charge, 73097 continued in fine form and, after some wonderfully noisy starts from all stations en route, we touched 82mph through Rossett. All those on board will never forget that run and congratulated the driver at Chester; he admitted he had never worked his fireman so hard – the said person just about having the energy to nod in agreement! For me, that was it and even though there was one further day of steam working between Shrewsbury and Chester and Birkenhead, I had had my fill of the area.

---

* As an aside, contained within Chris Magner's *Steam's Finest Hour* is a quote by one of my travelling companions, the late John Hobbs a.k.a. Prestatyn, who was aboard one of the specials: 'Birmingham Snow Hill, which was shortly to become a ghost station, would rise like a phoenix two generations later from an inglorious grave, which was sacrificed on the Beeching altar of WCML electrification and railway marginality.'

Notebook extract of my 1,300-mile, 700 of which were steam, sixty-three-hour marathon.

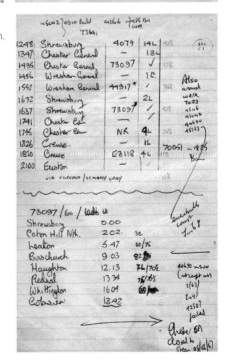

Logs as recorded by Joe Jolliffee
44917: 1445 Birkenhead Woodside to Paddington; load 6

Location	Time	Speed	Remarks
Gobowen	0.00		
Whittington	3.05	60½/75½	
Rednal	5.45	74½/71½	
Haughton Halt	6.57	79/82/78½	
Baschurch	10.05	80½/85	
Old Wood Halt	11.40	79	
Leaton	12.50	83½	
Coton Hill North	15.36		Brakes
Shrewsbury	17.18		(arr 1635)

73097: 1310 Paddington to Birkenhead Woodside: load 6

Location	Time	Speed	Remarks
Shrewsbury	0.00		(dep 1637)
Coton Hill North	2.03	38	
Leaton	5.50	60/75	
Baschurch	9.05	72/82	
Haughton Halt	12.22	76/70½	
Rednal	13.34	75/76½	
Whittington	16.04	68	
Gobowen	18.43		
Chirk	4.49	64/62	
Cefn	7.16	64	
Ruabon	9.38		
Johnston	3.09	45/55	
Wrexham	7.53		
Gresford	5.30	40	pws
Rossett	7.11	75½/82	

Pulford	8.23	80½/79/82	
		27/34½	sigs
Saltney Junction	13.33	30/37½	
			sigs
Chester	18.48		(arr 1744)

Out of the sixty different locos that I had travelled with on the line, with doggedness, patience, station hopping and extended periods of waiting, fifty-one required haulages were captured. In regard to the home allocations of Chester, Shrewsbury and Birkenhead, of the forty-four Stanier 5MTs, I missed out on just eight, with just one Birkenhead Stanier tank, 42548, not captured – although I did photograph her a long way from home at Newcastle in August 1964. I made my way home to 'the smoke', just missing, but not concerned about, the last Brit-worked 1605 Euston to Barrow at Crewe, with thoughts turning to the next 'last' event to be covered in the continuing decline of Britain's steam power.

Sunday the 5th saw service trains worked by 42587, 45116 and 44696, while tours saw 7029, 92203, 92234 and 44680 being used. The Unofficial Volunteer Birkenhead Steam Cleaning Gang, having adopted 42942 during 1966, also ensured 92203 and 92234 were dolled up for the specials.

Shrewsbury shed closed that weekend and Chester a month later, while the star of the Cambrian Coast Express, 75033, after initially joining five of her sisters in being put in store at Croes Newydd, eventually made her way to Carnforth and was withdrawn at the year's end. Birkenhead Woodside station closed on 4 November 1967, and the shed the following Monday.

So, in a flurry of steam activity, the curtain came down on direct services between Paddington and Chester, the service link that helped perpetuate the GWR's presence at Chester even beyond the boundary changes in 1958 that saw the London Midland claim dominance, although in reality the former GWR men ensured an element of continuity in their employment. As for the crews, despite working in far from ideal conditions, they entered into the spirit of the occasion and, on behalf of all steam enthusiasts who witnessed and participated in those far-off wonderful days, I would like to express collective thanks and appreciation for many happy memories.

RAILWAY WORLD

MAY 1967 VOL 28 NO 324

MARCH 4, 1967, will almost certainly go down as one of those special days in railway history which, unfortunately, are becoming all too common as steam traction declines on British Railways. Even five years ago the event which brought crowds to the lineside, the running of two " Castle " class locomotives, would have attracted little notice for it was part of the everyday railway scene. But today there are no Great Western steam locomotives left in normal service and trains hauled by the two surviving " Castle " class locomotives, *Pendennis* and *Clun*, within half an hour of each other over the same route between Banbury and Chester evoked such strong nostalgia as to bring observers from far and wide to throng the lineside and stations along the route. Indeed, the two excursions, 'organised by the publishers of this journal produced considerable revenue in fares and from platform tickets The large crowds at Birmingham Snow Hill and Chester had to be seen to be believed and it is clear that interest in the steam locomotive, now that it has gone from everyday service in most parts of the country, is greater than it has ever been. We have already referred in recent issues to the BRB's attitude towards the running of steam specials, and that, for the present, dispensation has been given to the running of steam locomotives on enthusiasts' specials in areas where steam locomotive facilities are available. What happens after the end of steam as a whole remains to be seen, since BR obviously is not going to maintain coal and watering facilities for diesel and electric traction. What is clearly needed is an arrangement between representatives of the principal railway societies and BR whereby strategic watering points on selected routes might be retained specially, with maintenance costs included in fares for future steam-hauled trains.

More serious, however, on March 4 and on other recent occasions, was the behaviour of too many so-called enthusiasts in their abuse to and disregard of railway staff and other enthusiasts in trying to achieve the best photographic vantage points. In some instances there were what can only be likened to rugby scrums, with elbows and feet well to the fore when photographers forced their way to the front of a crowd. The behaviour of a few in ignoring instructions by railway staff in crossing running lines in the path of oncoming trains was just reckless. That some enthusiasts were permitted to encroach along the lineside at stations was not a concession as a right, and it is only fair when rules are bent on such occasions that the instructions of railway staff and police are strictly followed. The action of those ignoring such advice was nothing short of disgraceful. It only needs one accident for the BRB to ban for good all enthusiasts' specials despite the interest and goodwill that they can generate towards the railways.

Editor: G. M. KICHENSIDE
al Subscription Rate including postage Home: 49s Overseas: 45s
Second class postage paid at New York post office, N.Y.

IAN ALLAN
TERMINAL HOUSE,
SHEPPERTON,
MIDDLESEX.
Walton-on-Thames 28484

Public interest in this occasion was on an almost incredible scale. The actual participation in it of nearly a thousand enthusiasts was a part only of that interest. Spread out over the entire route the number of spectators that turned out to see the " Castle " 4-6-0s with no exaggeration must have run into not far short of six figures. Snow Hill platforms at Birmingham were a seething mass of the populace, and there were very large crowds also at Banbury, Wolverhampton, Shrewsbury and Chester. Furthermore, over the whole route we were seldom out of sight of photographers, especially at points where the locomotives were likely to be working hard, as, for example, at the top of Hatton bank; several miles of film in the aggregate must have recorded their impressions that day of steam power still at work. It was indeed a memorable experience.

In these circumstances it is far from easy to understand the British Railways Board's reluctance to operate such enthusiasts' excursions as this one in the future. The opening up and staffing of disused branch lines, or the movement of special locomotives over long distances at the whim of excursion organisers, certainly is open to objection. In any event, also, with water-troughs now removed, steam locomotives can be operated only where water-cranes are still in existance, unless, like Alan Pegler's *Flying Scotsman*, they are equipped with a second tender carrying water only. On the credit side of the ledger, so far as concerned the BRB, fares and catering on these two specials must have brought in at least £3,000, and the locomotives themselves, in perfect running order, were supplied from outside sources.

Ex GWR Lines. The final days of the steam services on the Shrewsbury to Chester and Birkenhead line probably attracted more railway enthusiasts than any other event in railway history. The beginning of the end really began months ago when Chester, Shrewsbury and Birkenhead men began to get a keenness for providing out of the ordinary running. This carried on right through to the end as will be seen from later notes. Sunday 26/2 saw the first of several rail tours, when the Club 'Severn and Dee' rail tour headed northwards behind 7029 "Clun Castle." One point of interest with this engine is that it is now in a condition that never occurred — it has a double chimney and carries "Great Western" in full on the tender. This engine, of course, was never owned by GWR as it was not built until 1950 and is not therefore a correct preservation. A visit to Croes Newydd Shed on this date found 27 engines (23 steam and 4 diesels) on shed — 45130, 48122/252/87/440/632/97, 75002/46/52/60/71, 76040/8/88, D323/41/73 and D5143, together with the following either stored or withdrawn, 4646, 9641, 44776/872, 45198, 48325/85/669.

The following Saturday 4/3, was the last weekday service and the following is a list of the locos working the various trains :-

07.40 Birkenhead - Paddington	73141 9H to Chester	
08.55 „ „	42616 8H „ „	44690 10F Chester - Shrewsbury
11.45 „ „	73139 9H „ „	45116 6A „ „
14.35 Chester - Paddington		73097 9H „ „
14.45 Birkenhead „	42616 8H „ „	44917 6A „ „
16.30 „ „	42587 8H „ „	
20.55 „ „	42587 8H „ „	
00.15 Paddington - Birkenhead		73139 9H Chester - Birkenhead
08.20 „ - Chester	44690 10F Salop - Chester	
09.10 „ - Birkenhead	45042 6A „ „	[illegible] 8H „ „
12.10 „ „	44678 8F „ „	42616 8H „ „
13.10 „ - Chester	73097 9H „ „	
14.10 „ - Birkenhead	Diesel „ „	42587 8H „ „
04.10 Shrewsbury - Aberystwyth	75033 6D	
09.40 Aberystwyth - Shrewsbury	75033 6D	
14.30 Shrewsbury - Aberystwyth	75021 6D	
18.05 Aberystwyth - Shrewsbury	75021 6D	
1X.82 Paddington - Birkenhead Rail Tour	— 4079 "Pendennis Castle" Didcot - Chester and return.	
	73026 9K Chester - Birkenhead and return.	
1X.83 Paddington - Birkenhead Rail Tour	— 7029 "Clun Castle" Banbury - Chester and return.	
	73035 9H Chester - Birkenhead and return.	

SUNDAY, 5/3/67.

08.35 Birkenhead - Paddington	42587 8H Birkenhead - Chester
12.40 Chester - Paddington	45116 6A Chester - Shrewsbury
15.25 Birkenhead - Paddington	42587 8H Birkenhead - Chester (this engine then failed)
21.40 Birkenhead - Paddington	44690 10F Birkenhead - Chester
14.10 Paddington - Birkenhead	45116 6A Shrewsbury - Chester
	44690 10F Chester - Birkenhead
1Z65 SLS Rail Tour	7029 "Clun Castle" Tyseley - Birkenhead
	92234 8H Birkenhead - Chester and return
	44680 5B Birkenhead - Birmingham S.H.
1Z66 SLS Rail Tour	44680 5B Tyseley - Birkenhead
	92203 8H Birkenhead - Chester and return.
	7029 "Clun Castle" Birkenhead - Birmingham S.H.

45116 had the last steam working off Shrewsbury MPD. 75033 was cleaned up by a group of photographers in the platform at Aberystwyth. Wooden front number plate and 89C shedplate were fitted, as was a GW Type top lamp bracket, which carried the old "Cambrian Coast Express" headboard as far as Welshpool.

One member spent most of the last fortnight travelling from Birkenhead to Shrewsbury and there now follow details of some of the runs which amplify the spirit in which this service went out.

On 4/3 73139 (Load 8) reached 64 at Port Sunlight on the 07.20 ex Chester to Woodside (00.15 ex Paddington). 73141 (load 6) ran well on the 07.40 ex Woodside keeping time throughout. 42616 was specially cleaned for the 08.55 and Driver Lomax (8H) reached 52 mph before Bromborough; 64 was attained at Mollington but checks prevented a higher speed. 44690 ran the train to Shrewsbury and attained 77 after Baschurch, although the loco dropped ½ minute on time. 44690 worked the 11.34 ex Shrewsbury back to Chester and reached 76 before Haughton. Both trains were 6 coach formations. 45042 on the 09.10 ex London reached 82 mph and was in Chester to time. 73141 again ran well on the 13.45 to Woodside, and reached 65 mph before Bebington. 42616 now in the hands of M. Roberts (8H) on the 14.45 ex Woodside did very well to reach 62 before Hooton. The 6 coach train packed with enthusiasts ran the 5½ miles from Rock Ferry to Hooton in 7¾ mins. Chester was reached 1 min. early. Driver Bearnard (Shrewsbury) was in charge of the 15.31 to Shrewsbury, with 44917 specially turned out for the job. This train was the last "mile a minute" steam express and the last scheduled BR steam from Chester to Shrewsbury. Of the passenger complement, no less than 82 were enthusiasts, with 47 others. Wrexham was reached in 17 minutes (SC19) Ruabon in 18 (SC19) and Gobowen to Shrewsbury took 17 minutes 17 seconds. (SC18), with a max. of 86 after Leaton. The finest run of the whole year, however, was made by Driver Webb (Shrewsbury) on the 13.10 ex Paddington on 4/3. Let the log speak for itself:-

Engine 73097. Load 6 bogies.

Distance		Schedule	Actual	Speed
0	Shrewsbury	Nil	0.0	0
	Leaton		5.24	60
	Oldwoods		7.16	64
	Baschurch		9.01	75/76
	Haughton		12.19	82 Max.
	Rednal		13.32	76/77
	Whitington		16.00	73
			(signals severe to 5 mph)	
18.1	Gobowen	21	18.47 (Net 18½)	

Driver Webb showed no respect for the gradient and went over the summit from Shrewsbury at Leaton at the rate of 60 mph!! The Shrewsbury - Gobowen section is mainly uphill not down! Having left Gobowen on the 15.31 from Chester, one member arrived back exactly 38 minutes 15 seconds later, by catching 73097 on the 16.37 ex Shrewsbury. The total running time Gobowen - Shrewsbury - Gobowen was only 36.00 minutes for 36 miles 18 chains or less than "even" time. Saturday evening was the turn of the Birkenhead line to shine. 42587, load 7, on the 14.10 ex London ran the 8 miles to Hooton in 11 minutes 44 seconds (Sc.14), including 20 mph slack at Ledsham Junction to the slow line. 42587 cleared Capenhurst summit at 60 mph and the 1 mile 67 chains to Bromborough took 3 minutes 18 seconds with a 50 mph max. A special stop was made at Bebington (2 miles 4 chains) in 3 minutes 59 seconds (63 max). On the 20.55 ex Woodside, load 8 - 275 tons (including sleeping car) 42587 ran the slightly up grade section from Bebington to Bromborough in 4 minutes 38 seconds, and reached 50 mph again. (Sc. 6 minutes). This was a record-breaking run. The 23.08 ex Chester ran very well and kept good time. The wonderful running continued on the Sunday 5/3. 42587 ran before time on the 08.35 ex Woodside to Chester. D1712 on the Gobowen-Shrewsbury section covered the distance in 15 minutes 39 seconds. 7029 "Clun Castle" on the SLS Special reached 77 before Haughton. 92234 worked the special from Birkenhead to Chester and with no effort whipped the 8 coaches to 72 at Mollington. 92203 on the following special reached 68 mph. 7029 returned to Birmingham and reached 73 at Mollington. 45116 had the honour of the last steam workings on the line (12.40 ex Chester and 18.05 return) in lieu of a diesel and gave a wonderful firework display coming out of Shrewsbury on the return run. 44690 working the last Paddington train (the 21.40 ex Woodside) attained 70 down Mollington bank. D297 was in charge of the last train ex Paddington (16.10 ex London). A special headboard beautifully made and inscribed "GWR BIRKENHEAD - PADDINGTON" 1861 - 1967, was carried by 45116, 44690 and 42587. The headboard included the GWR crest. In

And Finally: The Last Weekend

I couldn't have finalised this book on a better note. Yet another memorable day in the life of a 1960s steam chaser. It didn't matter whether it was climbing Talerddig behind a struggling Standard 4MT or being present during the two-coach final day shuttle on the K1-operated Alnwick branch. Having caught my final required Jubilee and Brit and been aboard the last Trans-Pennine tank-hauled train into Bradford, there had been many 'lasts' until then and there would be many more to come. Ahead of me lay the final steam train into London, the last steam-operated Yorkshire Pullman and Belfast Boat Express trains and, of course, the very last public steam train in Britain into Liverpool Exchange. Unbelievable and unrepeatable occasions of which I am fortunate enough to be able to say, 'I was there!'

Glossary of Terms

BR	British Rail(ways) (1948–1997)
BRB	British Railways Board
BRUTEs	British Rail Universal Trolley Equipment
DEMU	Diesel–Electric Multiple Units
DL	Diesel Locomotive
DMU	Diesel (mechanical) Multiple Units
ECML	East Coast Main Line
ECS	Empty Coaching Stock
EMU	Electric Multiple Unit
GCR	Great Central Railway (1897–1922)
GWML	Great Western Main Line
GWR	Great Western Railway (1835–1947)
K&ESR	Kent & East Sussex Railway
L&SWR	London & South Western Railway (1838–1922)
LMR	London Midland Region of BR (1948–1992)
LMS	London Midland & Scottish Railway (1923–1947)
M&SWJR	Midland & South Western Junction Railway (1884–1891)
MNA	Master Neverers Association
S&D	Somerset & Dorset Railway (1862–1922)
SO	Saturdays Only
SR	Southern Region of BR (1948–1992)
SX	Saturdays Excepted (Mondays to Fridays)
TPO	Travelling Post Office
WCML	West Coast Main Line
WR	Western Region of BR (1948–1992)

Appendix I

Western Region
Steam Allocations
(Month Ending) 1965–66

CLASS	Jan	Feb	Mar	Apr	May	Jun	Jul	Aug	Sep	Oct	Nov	Dec	Jan	Feb	Mar
14xx	2	2	2	2											
16xx	9	9	8	8	8	7	5	4	4						
2251	14	13	12	12	2										
2884	25	24	23	21	19	16	4	3	3	1					
57xx	124	117	112	106	104	68	47	39	27	20	14	2	2	2	
5101	16	16	16	16	14	5	5	5	3	3					
42xx	8	6	6	2	2	1	1								
Hall	29	26	24	22	22	20	19	16	15	19	7				
Castle	4	4	4	4	4	1	1	1	1	1	1				
5205	10	10	6	6	4	2	1	1							
56xx	31	30	26	25	16	5	3	2	2	1					
61xx	30	30	28	28	28	24	24	23	20	15	12				
Grange	26	23	22	21	20	19	15	14	12	11	4				
M Hall	35	32	28	27	26	24	22	19	18	17	16				
72xx	9	8	8	5	4										
74xx	2	2	1												
Manor	8	8	8	7	6	6	5	5	3	3	2				
94xx	32	32	32	30	22										

412xx	14	14	14	14	13	13	9	9	9	9	9	9	9	8
Fow 4F	1	1	1	1	1									
LMS 4F	7	7	7	6	5	3	3	3	2	2	1			
LMS 3F	3	3	3	3	3	3	3	3	3	3	3	2	2	2
LMS 8F	9	9	8	8	7	7	6	6	6	5	4	4	4	3
73xxx	18	18	16	13	11	11	11	5	4	4	4			
75xxx	9	9	9	8	8	8	8	8	8	7	6			
78xxx	3	3	3	3	3	3	3	3	3	3	2			
80xxx	8	8	8	7	7	6	6	6	5	5	4	4	3	3
82xxx	12	12	12	12	12	12	9	5	5	4	3			
92xxx	23	23	21	20	18	18	15	12	9	8	7			
TOTAL	521	499	468	437	389	282	225	192	162	141	99	21	20	18

227

Appendix II

Western Region Sheds Open to Steam on 1 January 1965

Date closed	Shed	Running Total
31/12/64		23
01/65	Reading (81D)	22
03/65	Aberdare (88J), Treherbert (88F), Old Oak Common (81A)	19
04/65	Rhymney (88D)	18
05/65	Pontypool Road (86G)	17
06/65	Didcot (81E), Exmouth Junction (83D), Yeovil (83E), Neath (87A)	14
07/65	Cardiff Radyr (88B)	12
08/65	Cardiff East Dock (88A)	11
09/65	Westbury (83C)	10
10/65	Newport Ebbw Junction (86B), Llanelly (87F)	8
11/65	Bristol Barrow Road (82E)	7
12/65	Worcester (85A), Severn Tunnel Junction (86E)	5
01/66	Gloucester Horton Road (85B), Oxford (81F), Southall (81C)	2
03/66	Bath Green Park (82F), Templecombe (83G)	0

Appendix III

Ex Western Region Sheds, Now LMR, Open to Steam on 1 January 1965

Date closed	Shed	LMR	Running Total
31/12/64			9
01/65	Oswestry (89D)	6E	8
06/65	Leamington (84D)	2L	7
07/66	Stourbridge Junction (84F)	2C	6
10/66	Banbury (84C)	2D	5
11/66	Tyseley (84E)	2A	4
12/66	Machynlleth (89C)	6F	3
03/67	Shrewsbury (89A), Wolverhampton Oxley (84B)	6D 2B	1
06/67	Croes Newydd (89B)	6C	0

Appendix IV

Preserved Manors

Loco	Name	Withdrawn	Preserved
7802	*Bradley Manor*	11/65 @ 6D	Severn Valley Railway
7808	*Cookham Manor*	12/65 @ 85B	Didcot Railway Centre
7812	*Erlestoke Manor*	11/65 @ 6D	Severn Valley Railway
7819	*Hinton Manor*	11/65 @ 6D	Severn Valley Railway
7820	*Dinmore Manor*	11/65 @ 6D	Gloucester & Warwickshire Railway
7821	*Ditcheat Manor*	11/65 @ 6D	West Somerset Railway
7822	*Foxcote Manor*	11/65 @ 6D	Llangollen Railway
7827	*Lydham Manor*	10/65 @ 6D	Dartmouth Steam Railway
7828	*Odney Manor*	10/65 @ 6D	West Somerset Railway

Appendix V

Great Western Locomotives (Month Ending) 1965–66

The Great Western Steam Retreat

CLASS	Jan	Feb	Mar	Apr	May	Jun	Jul	Aug	Sep	Oct	Nov	Dec
14xx	2	2	2	2								
16xx	13	13	12	11	11	10	8	7	7	3	3	3
2251	14	13	12	12	2							
2884	32	30	28	26	23	20	7	3	3	1		
57xx	166	159	154	146	142	105	83	75	60	50	39	27
5101	28	27	27	26	24	13	13	13	8	3		
42xx	8	6	6	2	2	1	1					
Hall	44	39	37	31	31	28	27	23	20	9	9	
Castle	12	4	4	4	4	1	1	1	1	1	1	
5205	10	10	6	6	4	2	1	1				
56xx	55	54	50	49	40	28	23	22	19	13	3	2
61xx	30	30	28	28	28	24	24	23	20	15	12	
Grange	45	42	41	40	38	35	31	29	23	11	4	
M Hall	43	39	35	34	33	31	29	25	23	17	16	
72xx	9	8	8	5	4							
74xx	2	2	1									
Manor	19	18	18	16	15	15	13	13	11	9	2	
81xx	1	1	1	1	1							
94xx	32	32	32	30	22							
TOTAL	565	529	502	469	424	313	261	235	195	132	89	32

Appendices

Jan	Feb	Mar	Apr	May	Jun	Jul	Aug	Sep	Oct	Nov	Dec
3	2	2	2	2	2	2	1				
25	25	23	20	20	20	13	11	8	5	3	
2	2	2	2								
30	29	27	24	22	22	15	12	8	5	3	

Appendix VI

This Train (No Longer) Terminates Here

A summary of closed lines (to passengers) visited during my travels

Train service	Closure	Visited	Traction	Notes
Kemble/Tetbury	Apr 64	Mar 64	DMU	
Kemble/Cirencester	Apr 64	Mar 64	DMU	
Taunton/Yeovil Town	Jun 64	Apr 64	Steam	
Neath/Pontypool Road	Jun 64	Mar 64	Steam	
Worcester/Bromyard	Sep 64	Sep 64	DMU	Closed a week later
Hereford/Gloucester	Nov 64	Sep 64	Steam	
Morfa Mawddach/Ruabon	Jan 65	Nov 64	Steam	Part closed (flooding) Dec 64
Welshpool/Whitchurch	Jan 65	Nov 64	Steam	
Llanymynech/Llanfyllin	Jan 65	Nov 64	Steam	
Staines West/West Drayton	Mar 65	May 64	DMU	
Halwill/Torrington	Mar 65	Jul 64	Steam	
Chippenham/Calne	Sep 65	Sep 65	DMU	Last day
Torrington/Barnstaple	Oct 65	Jul 64	Steam	

Axminster/Lyme Regis	Nov 65	Jul 64	DMU	
Poole/Bristol	Mar 66	Feb 66	Steam	Closed a week later
Evercreech Jn/Highbridge	Mar 66	Sep 65	Steam	Closed a week later
Seaton/Seaton Junction	Mar 66	Jul 64	DMU	
Barnstaple/Taunton	Oct 66	Jul 64	Steam	
Okehampton/Wadebridge	Oct 66	Jul 64	Steam	
Halwill/Bude	Oct 66	Jul 64	Steam	
Yeovil Town/Junction	Oct 66	Jul 64	Steam	
Padstow/Bodmin Road	Jan 67	Jul 64	DL	
Exmouth/Sidmouth Jn	Mar 67	Jul 64	Steam	
Barnstaple/Ilfracombe	Oct 70	Jul 64	DL	
Coleford Jn/Okehampton	Jun 72	Jul 64	Steam	
Maiden Newton/Bridport	May 75	Jan 67	Steam	Rail tour

Sources

Books:

British Rail Main Line – Gradient Profiles (Shepperton: Ian Allan, 1966).

Fuller, Aidan L.F., *British Locomotive Shed Directory* (several editions).

Longworth, Hugh, *British Railways Steam Locomotives 1948–1968* (Manchester: Crecy, 2005).

Magner, Chris, *Steam's Finest Hours* (self-published, 2018).

Whittaker, Nicholas, *Platform Souls* (London: Icon Books Ltd, 2015).

Other:

BR Database (www.brdatabase.info).

Locomotive Club of Great Britain Monthly Bulletin magazine.

Railway World magazine.

Six Bells Junction website (www.sixbellsjunction.co.uk).

By the same author

THE GREAT
STEAM
CHASE

The Last Days of Steam on
BR'S SOUTHERN REGION

KEITH WIDDOWSON

978 0 7524 7957 6

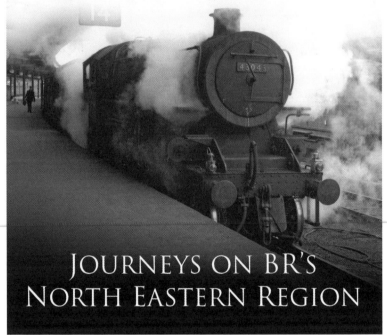

KEITH WIDDOWSON

RIDING YORKSHIRE'S FINAL STEAM TRAINS

JOURNEYS ON BR'S NORTH EASTERN REGION

978 0 7509 6047 2

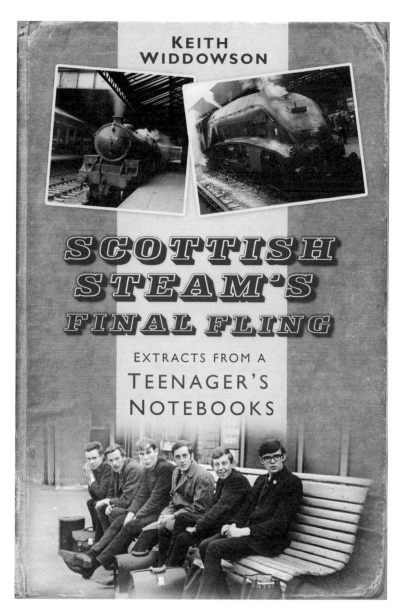

KEITH WIDDOWSON

SCOTTISH STEAM'S FINAL FLING

EXTRACTS FROM A TEENAGER'S NOTEBOOKS

978 0 7509 7022 8

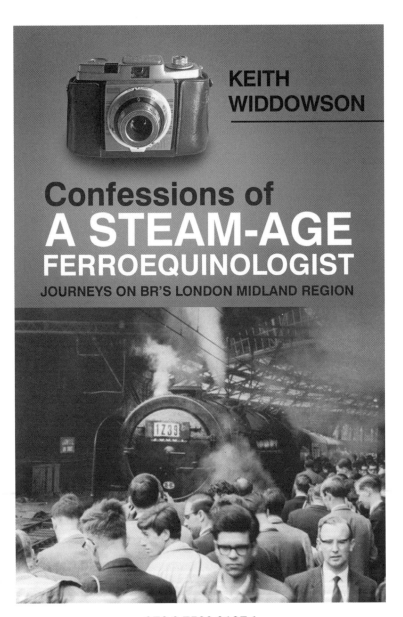

KEITH
WIDDOWSON

Confessions of
A STEAM-AGE
FERROEQUINOLOGIST
JOURNEYS ON BR'S LONDON MIDLAND REGION

978 0 7509 9197 1

The destination for history
www.thehistorypress.co.uk